D0561042

WITHDRAWN

THE BIG THAW

THE BIG THAW

Travels in the Melting North

ED STRUZIK

Foreword by Gerald Butts,
President & CEO, WWF-Canada

John Wiley & Sons Canada, Ltd.

Portions of this book were previously published in the *Toronto Star* and the *Edmonton Journal.*

Library and Archives Canada Cataloguing in Publication Data
Struzik, Ed, 1954–
The big thaw : travels in the melting north / Ed Struzik.

Includes index.
ISBN 978-0-470-15728-2

1. Climatic changes—Environmental aspects—Arctic regions. 2. Habitat (Ecology)—Arctic regions. 3. Global warming. 4. Global environmental change. 5. Arctic regions—Environmental conditions. 6. Arctic regions—Climate. 7. Inuit—Social conditions. 8. Arctic regions—Politics and government. I. Title.

QC994.8.S82 2009 363.738'7409113 C2009-900729-0

This book is printed with biodegradable vegetable-based inks. Text pages are printed on 55lb 100% post-consumer waste (PCW) paper and are Forest Stewardship Council (FSC) certified; the insert is printed on 80lb gloss, FSC-certified paper; the endpapers are 30% PCW and FSC-certified; the jacket contains 10% PCW; and the case material is 10% PCW and is FSC-certified.

Production Credits
Cover design: Ian Koo
Cover photo and photo inserts: Ed Stuzik
Interior design and typesetting: Adrian So
Printer: Friesens

John Wiley & Sons Canada, Ltd.
6045 Freemont Blvd.
Mississauga, ON
L5R 4J3

Printed in Canada

1 2 3 4 5 FP 13 12 11 10 09

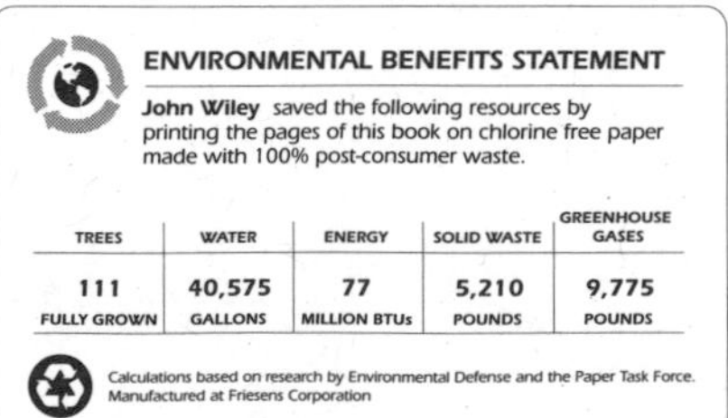
ENVIRONMENTAL BENEFITS STATEMENT

John Wiley saved the following resources by printing the pages of this book on chlorine free paper made with 100% post-consumer waste.

TREES	WATER	ENERGY	SOLID WASTE	GREENHOUSE GASES
111	40,575	77	5,210	9,775
FULLY GROWN	GALLONS	MILLION BTUs	POUNDS	POUNDS

Calculations based on research by Environmental Defense and the Paper Task Force. Manufactured at Friesens Corporation

CONTENTS

ARCTIC OCEAN
QUEEN ELIZABETH ISLANDS
Ellesmere Island
Axel Heiberg Island
Eureka
Fossil forest
Barrow
Beaufort Sea
McClure Strait
Melville Island
Bathurst Island
Haughton Crater
Polar shelf
Banks Island
Arctic National Wildlife Refuge
Ivvavik National Park
Herschel Island
Alaska (UNITED STATES)
Yukon River
Tuktoyaktuk
Fairbanks
Old Crow
Aklavik
Inuvik
M'Clintock Channel
Delta Junction
Victoria Island
Great Bear Lake
Mackenzie River
Yukon
Kluane National Park
Mile 1004 (Location of "Squirrel Camp")
Back River
Nunavut
Goathead Mountain
Alaska Highway
Brintnell Glacier
Northwest Territories
Whitehorse
Yellowknife
Rankin Inlet
Great Slave Lake
PACIFIC OCEAN
Lake Athabasca
Churchill
British Columbia
Alberta
Saskatchewan
Manitoba

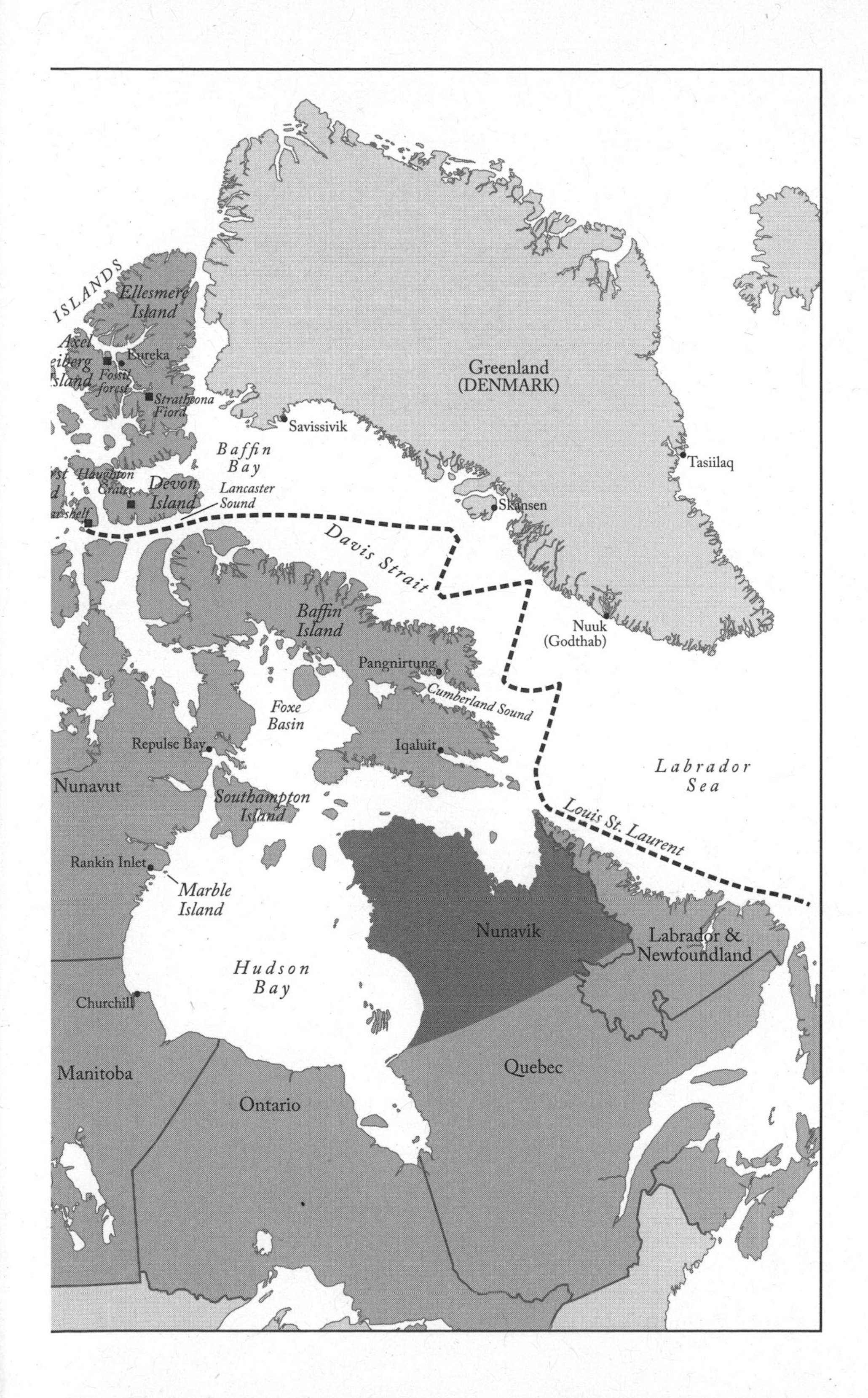

ISLANDS
Ellesmere Island
Axel Heiberg Island
Eureka
Fossil forest
Strathcona Fiord
Haughton Crater
Devon Island
Lancaster Sound
Baffin Bay
Savissivik
Greenland
(DENMARK)
Tasiilaq
Skansen
Davis Strait
Baffin Island
Nuuk
(Godthab)
Pangnirtung
Cumberland Sound
Foxe Basin
Repulse Bay
Iqaluit
Labrador Sea
Nunavut
Southampton Island
Louis St. Laurent
Rankin Inlet
Marble Island
Nunavik
Labrador &
Newfoundland
Hudson Bay
Churchill
Manitoba
Quebec
Ontario

FOREWORD

"It is one thing to be traveling in ice fog and quite another to describe it to someone who has never experienced the surreal feeling of being in an icy tomb such as this. No amount of words can adequately explain how limited vision, the total lack of smell, the stinging sensation on the skin and the amplification of sounds are all part of the weird experience."

Beyond being a vivid depiction of a particular personal experience in the High Arctic, the above passage from *The Big Thaw* is an apt description of the sensory deprivation with which we, at the top of the food chain, appear to be navigating this swiftly tilting planet of ours. While too much of the world spent the last decade debating whether climate change is real, people at its northern reaches turned their attention to living with its consequences.

We should all be thankful that Ed Struzik, with this marvelous book, turns in such an astonishing rebuttal to his own supposition.

I met Ed for the first time on a canoe trip down the Nahanni almost 10 years ago. You get to know people better after a week in the wilderness than a decade in the city. What struck and has stayed with me about Ed was his uncanny ability to soak up the truth about his environment and the people within it. In the intervening years, whether traveling up north or speaking with southerners engaged in northern issues, I have been equally struck by the universal esteem in which Ed is held. More than once, I've seen that respect for Ed by people who agree about nothing else!

There is a particularly poignant story—in this book of so many wonderful stories—set in Repulse Bay. The author is on polar bear and narwhal watch, partnered with a young Inuk named Jonah Siusangnark. Siusangnark is reserved at first, probably suspicious of the motives of another southerner who has come north to ply his trade. Over many frigid nights, Struzik earns his trust by engaging him in conversation, asking questions about his life, his community's history and absorbing his perspective on the many challenges of life in the Arctic.

Most of all, he gains that trust by being there. And staying there.

Toward the end of their time at camp, they are rewarded for their efforts with a glorious full moon. They talk about how cold and dead the lunar surface must be, and Siusangnark makes a common-sense query: if it's even colder there than here and nothing lives there, "Then why would anyone want to go there when they could come here?"

Good question. Even in my country, Canada, where it composes a third of our land-mass and half of our coastline, the Arctic is barely there. Throughout the western world, it is mythological, present in our collective imagination, but rarely seen as it is.

Struzik makes plain that we *Kabloona* are increasingly present in this majestic place, if not in person than via some of our most uncomely ambassadors. From the southern toxins that now infect Inuit and Inuvialuit mothers' milk to the northward migration of diseases that precipitate massive wildlife die-offs, we are there and our presence is decisive. The full weight of southern civilization is just now beginning to be felt because of the literal and figurative shift in the Arctic landscape wrought by the onset of climate change. This is, of course, the main-stage for the compelling drama Struzik depicts in *The Big Thaw*.

The great American author William Faulkner once wrote, "The past isn't dead. It isn't even past." Nowhere on Earth is this truer today than in the Arctic. Three hundred years of our history are stirring to life there as a new force of nature. This force will shape the collective future of our species.

The people of the Arctic will be the first to bear the consequences. From an open Northwest Passage to the gathering energy gold rush in the Beaufort; from an uncertain future for polar bear and other Arctic wildlife to the real threat of conflict over irreconcilable sovereignty claims; it is all here in *The Big Thaw's* evocative, sympathetic and painstakingly researched narrative.

Ultimately, it is a human story that Ed Struzik describes for us. For those of us, in particular, who have benefitted most from the fruits and comforts of modern life, it is our story. The future of the Arctic is our future.

— Gerald Butts, President & CEO, WWF-Canada

INTRODUCTION

Many countries—and they are to be envied—possess in one direction or another a window which opens out on to the infinite—on to the potential future. . . . The North is always there like a presence, it is the background of the picture, without which Canada would not be Canadian.

—French geographer André Siegfried, 1937

In the summer of 1985, helicopter pilot Paul Tudge was flying over Axel Heiberg Island in the High Arctic when he spotted what he thought were tree stumps sticking out of the ground near the edge of a giant ice cap. When Tudge reported the sighting to scientists, they were initially skeptical. Nevertheless, geologist James Basinger and Australian paleobotanist Jane Francis flew up the next year to have a closer look. It didn't take long for them to realize that they had found the Holy Grail of Arctic fossil forests 1,200 miles (2,000 km) from the nearest stand of trees.

Not only were there tree stumps sticking out of the permafrost, but some that the scientists eventually found buried below were more than 8 feet (2.5 m) wide and more than 16 feet (5 m) long. Through geological detective work, Basinger and his colleagues were able to date these trees to a time 45 million years ago. Many of the nuts, seeds and cones were so perfectly preserved they looked as if they had only recently fallen to the ground. Moreover, the specimens were mummified, not petrified. Some still held the sap they oozed before a catastrophic flood buried them in an anaerobic tomb of sand.

Once news of the discovery got out, first in a scientific journal and then in *Time* magazine, pseudo-scientists who had probably seen the movie *Jurassic Park* one too many times asked Basinger for samples of the golden amber in the hopes they might extract the DNA of an insect that may have got trapped inside. Creationists, on the other hand, were convinced Basinger had found the mountaintop where Noah's Ark had landed. Appalled by this lunacy, Basinger stayed out of the media spotlight for many years before allowing me to join him at the site in the summer of 1998.

By that time, Basinger and Francis had assembled a picture of the scene as it must have been so many million years ago: a Dawn Redwood swamp filled with royal ferns and cypress that flourished downstream of an upland environment dotted by pine, spruce and walnut. Not far away, colleagues Jaelyn Eberle and John Storer had unearthed the scant remains of a brontothere, a rhinoceros-like animal that lived in this world that was as warm and lush as today's Carolinian forests of Georgia in the United States. The mean ocean surface temperatures back then were between 50°Fahrenheit and 59°Fahrenheit (10°C and 15°C), a far cry from the 14°Fahrenheit (-10°C) experienced today.

This wasn't the first time scientists had evidence of an ancient Arctic Eden, and it wouldn't be the last. Most of the world, however, barely noticed when Mary Dawson and Robert West, vertebrate paleontologists at the Carnegie Museum of Natural History and the Milwaukee Public Museum respectively, excavated a rich vein of varied life forms in the High Arctic in 1975. In among the rocks, gravel and peat along the icy shores of Strathcona Fiord on Ellesmere Island, they found fossil fragments of alligators, giant tortoises, snakes, lizards, tapirs, hippo and rhino-like animals that lived 55 million years ago.

Nor did anyone but a handful of paleontologists and geologists pay much attention in the 1980s when Richard Grieve and scientists from Canada, the United States and Germany unearthed, among other fossils, a primitive rhinoceros at Haughton Crater on Devon Island that turned out to be 39 million years old. While not quite as warm as it was 45 million or 55 million years ago, the mean annual temperatures there were high enough to sustain a mixed conifer-hardwood forest in the years before a meteorite speeding at 37,300 miles per hour (60,000 km/h) slammed into the ground with a force of energy equal to that of a thousand-megaton blast. When the white-hot remnants of jagged breccia rock fell to earth and mixed with ground and surface water, they spewed a deadly cloud of steam and gas that killed all life within a 90-mile (150 km) radius.

As an Arctic traveler and science writer who had visited both the Axel Heiberg fossil forest and Devon Island crater sites, the climatic implications of these findings were still not entirely clear to me even then. Not until I traveled to Strathcona Fiord in 1999 did I come to realize that since the deep freeze of Snowball Earth[1] ended 630 million to 750 million years ago, the Arctic has often been anything but the forbiddingly cold place we know so well today. There, on a hilltop at Strathcona Fiord just 6 miles (10 km) from Mary Dawson's old camp, Richard Harington, a paleontologist with the Canadian Museum of Nature, was in the last stages of excavating a 4.5-million-year-old beaver pond site that had been found years earlier by geologist John Fyles. In addition to the remains of miniature beavers that would have been constantly on the lookout for predators such as ancestral black bears, weasel-like carnivores and Eurasian badgers, Harington and his colleagues had unearthed the fossils of three-toed horses and deer that had fangs but no antlers. Some of the fossils were so exquisite the excavation team was able to determine what tundra bunnies were eating at the time. Temperatures 4.5 million years ago were at least 10 degrees warmer in summer and 15 degrees warmer in winter than they are today.

Harington's extraordinary discoveries at Strathcona Fiord followed his equally remarkable findings along the Whitestone River in the

1. Snowball Earth is the term scientist Joe Kirschvink (California Institute of Technology) coined in the 1980s to describe global glaciation, the time in the Earth's history when most if not all of the world was covered in snow and ice. The dates of this glaciation are still being debated.

northern Yukon. There, he and Peter Lord, a Gwich'in from Old Crow in the northern Yukon, unearthed the fossils of woolly mammoths, giant sloths and 6-foot-tall (2 m tall) beavers that shared the Beringian world of the Yukon and Alaska with scimitar cats, American camels and mastodons between 70,000 and 9,000 years ago.

No one really knows exactly why it was warm for so long in the Arctic. Unlike our current situation, where greenhouse gas emissions are the main driver for climate change, the Earth's climate has responded in the past to variations in the Earth's eccentricity, axial tilt and precession.[2] Volcanic ash, the formation of gyres,[3] methane seeping from permafrost and percolating from the ocean floor were other factors that likely controlled the Earth's thermostat.

What we do know from fossil evidence found in ancient lake and ocean floor beds, ice cores and permafrost suggests that a trend to cooling began shortly after large mammals replaced the dinosaurs 65 million years ago. This gradual cooling, interspersed with episodes of increasing warmth, led to the gradual buildup of ice sheets in the Arctic shortly after the meteorite slammed into Devon Island. By the time Harington's miniature beavers were watching out for Eurasian badgers on Ellesmere Island, a catastrophic cycling of advancing and retreating glaciers in the Arctic began to take hold.

In relatively rapid-fire fashion, the cold wiped out the Arctic forests, the miniature beavers and the Eurasian badgers. Even the woolly mammoths, American mastodons, sabertoothed cats and giant sloths that took their place hundreds of thousands of years later were unable to weather the cycles of cooling and warming that glaciated 30 percent of the Earth's surface at one end of the extreme and turned huge swaths of tundra into forests and shrublands on the other. What we are left with today in the Arctic are the survivors—the polar bears, narwhals, bowheads, belugas,

---

2. Eccentricity is the shape of the Earth's orbit around the sun. Axial tilt is the inclination of the Earth's axis in relation to its plane of orbit around the sun. Precession is the Earth's slow wobble as it spins on its axis.

3. Gyres are giant circular ocean surface currents that swirl both clockwise and counterclockwise.

muskoxen and Peary caribou—that were able to evolve and adapt to this vicious cycling and to the deep freeze that has characterized the Arctic's most recent past.

These are heady times for scientists. Now that man-made greenhouse gases are rapidly warming the Arctic, many scientists fear that we have reached the tipping point, the term climatologist Mark Serreze uses to describe what happens when a critical threshold of warming in the Arctic passes the point of no return.

With sea ice melting, glaciers receding and Arctic storms picking up steam, the Arctic is moving toward a new state unlike anything recorded in human history. No one knows exactly what to expect in the future, but if the past tells us anything, the consequences will be catastrophic in many ways.

A warmer and shorter ice season means less time for polar bears to hunt seals and more time for mosquitoes and blackflies to take their toll on caribou, muskoxen and nesting birds. Beluga whales, which hide under the ice to avoid killer whales, could also be threatened.

Heat threatens Arctic species in other ways as well. There's evidence that arctic char, arctic fox, narwhal and other northern species may not be able to compete if Pacific salmon, red fox and killer whales continue to migrate north into their territory. This is not just futuristic speculation. Increasing numbers of Pacific salmon have been showing up in Inuit nets. The ice-choked channels that used to stop or at least deter killer whales from moving into narwhal and beluga territory have opened up in the summer months. In 2006 and 2007, killer whales may have been responsible for chasing hundreds of beluga whales into the shallow Husky Lakes region of the western Arctic, where winter ice eventually trapped and killed them.

With so much ice melting in the Arctic, many more Alaskan polar bears are now denning on land rather than on the sea ice, and gray whales, which traditionally leave the Chukchi Sea for warmer waters along the California and Mexico coasts between October and mid-December, are beginning to stay north for the winter.

Human activity has also been facilitating this northward movement of southern animals. White-tailed deer are now exploiting the vast network of roads, cutlines[4] and pipeline right-of-ways carved through the thick boreal forest to the edges of the tundra and high alpine environments. Cougars, coyotes and other predators are following them into this new warmer world. And now that cold winters are no longer so murderous, diseases that were once unable to take root in the Arctic are beginning to move in—in some cases with those animals heading north.

Theoretically, a polar meltdown could shut down the ocean conveyor belt that brings warm water into the North Atlantic and moderates the climate of Great Britain and northern Europe. Rising sea levels brought on by this meltdown could displace the 104 million people who live in coastal areas that are within 3 feet (1 m) of the ocean surface. Those who live on higher ground also won't escape the coming changes. Polar ice is the genesis of cold fronts that bring rain and snow to much of the world. If it shrinks, winters won't go away. But the problems that people in drought- and wet-weather-stricken regions are now facing could get worse.

The rest of the world will also be vulnerable to boreal forest fires that will inevitably escalate in size and severity in the coming decades as the North heats up and lightning strikes increase. Few people in southern Canada and the northeastern areas of the United States realize it, but part of the suffocating smog they suffered through in the summer of 2004 contained fallout from massive fires burning in Alaska and the Yukon. All told, 5 percent of Alaska, an area equivalent to the states of Massachusetts and New Hampshire, and 4 percent of the Yukon burned that year.[5]

The future is not necessarily all filled with doom and gloom. Moist Arctic air could also bring relief to some drought-stricken areas. And some Arctic animals—such as the barren ground grizzly—will probably do well in a world in which it does not have to hibernate for so long.

---

4. Cutlines are straight, narrow bands through the forest cleared of trees, usually for the purposes of powerlines or seismic exploration for oil or gas.

5. Gabrielle Pfister and a team from the National Center for Atmospheric Research in Boulder, Colorado, estimated that the fires generated 66 billion tonnes of carbon monoxide, roughly the equivalent to all that produced by humans on the continental United States in that time.

Environmentalists would be horrified by the prospect, but a seasonally ice-free Arctic also holds out the very real possibility that a fortune in so far unexploited resources will soon find a way to market. The Northwest Passage, which has been too dangerous for ships to transport these resources, could be seasonally ice-free sometime within the next decade.

The development of the Arctic's vast oil, gas and mineral reserves would undoubtedly be a boon to a world economy starved for new, politically stable sources of fossil fuels and metals. A warmer Arctic, however, also opens the continent's back door to drug smugglers, illegal aliens, terrorists and energy-starved countries in desperate need of new sources of fossil fuels. Commercial shipping in the Arctic also raises the potential for an environmental disaster that could make the *Exxon Valdez*, the biggest man-made environmental disaster in North American history, look like a minor oil spill.

Mark Serreze cautions optimists who think there's plenty of time for humans to adjust or capitalize on the changes taking place. So far, he notes, the climate models that he and others at the National Snow and Ice Data Center at the University of Colorado, Boulder, have put together have not been wrong, but they have seriously underestimated how quickly the changes that have happened already would occur. The 3,860-square-mile (10,000 sq. km) annual ice-cover losses that he and other climatologists used to talk about have turned into 30,888 square miles (80,000 sq. km) or more.

"It's not what we know, it's what we don't know," he said. "The paleoclimate record tells us that the system can change very, very quickly, on the order of just ten years. I suspect that there are surprises ahead that we won't be ready for."

The recent history of the Arctic is filled with these surprises. Greenland was warm enough for Erik the Red to establish a colony on the island in AD 985. For nearly five hundred years, the Viking settlers farmed, fished and harvested birds and marine mammals. Then the so-called Little Ice Age thrust the island into a deep freeze. In less than a generation's time, the Greenland colony collapsed and disappeared.

Archeological evidence suggests that the Inuit in Canada suffered a similar ordeal. But unlike the Vikings of Greenland, the Inuit were nomadic. Archeological sites such as the one Bryon Gordon excavated on Nadlok, an island on the Burnside River in the Canadian Arctic, show that some Inuit moved far inland during the Little Ice Age, successfully shifting their hunting strategies away from polar bears, arctic char, whales and seals to caribou, lake trout and muskoxen. Some of them used caribou antlers and hides to frame summer and fall dwellings that would have traditionally been supported by whalebone or driftwood. They even traded with their mortal enemy—the Dene—in order to survive. When the Little Ice Age ended, they returned to the coast to continue their former way of life.

The scale of some of these feast-and-famine episodes is truly breathtaking. On a single hilltop at the north end of Banks Island in the High Arctic, there are the remains of 581 muskox skulls, 29 food caches and 17 tent rings that suggested there were once thousands of muskoxen and many people living on the island. But by the time the Little Ice Age ended sometime around 1850, the animals were virtually all gone and so were most of the Inuit. A live animal wasn't spotted until 1952, when two Canadian scientists thought they saw a single bull during their three-month-long survey of the island. Now that the heat has returned to the Arctic in a big way, muskoxen are once again thriving. As many as 64,000 animals now live on the island.[6]

Like muskoxen, most Arctic animals have evolved in a way that allows them to adapt to these climate cycles. The situation is much different now. The climate is moving toward unidirectional change instead of the cycle that led to feast and famine and boom and bust. And as Serreze points out, it is moving very fast. The ecosystem that supports polar bears, narwhal, hooded seals and other Arctic animals that rely on sea ice is on the verge of collapse. And there is now every reason to believe that habitat loss will be long-term—if not permanent.

---

6. An alternative theory suggests the presence of an abandoned ship, the *Investigator*, which had been used to search for the lost Franklin expedition (1850–53), attracted so many Inuit to the north end of Banks Island that they eventually depleted the muskoxen population.

Up until the summer of 2007, there were still some scientists who were skeptical that the climate system that keeps the Arctic Ocean cold and the ice frozen for much of the year could unravel so quickly. But in 2007, the ice retreated so far beyond all expectations that most experts were shocked, if not stunned, by what they saw in the satellite imagery. Across the Arctic as a whole, the meltdown was where climatologists expected it would be in 2030. What really made the big melt of 2007 an eye-popping experience was the absence of ice in areas where it almost never melts. The so-called "mortuary" of old ice that normally chokes M'Clintock Channel in the High Arctic was almost all gone. What's more, Viscount Melville Sound, "the birthplace" of a great deal of new Arctic ice, was down to half its normal summer cover. "The ice is no longer growing or getting old," noted John Falkingham, chief forecaster for the Canadian Ice Service. "Ten years from now," he said that fall, "we may look back on 2007 and say that was the year we passed the tipping point."

Shortly before I set off in the spring of 2006 on a series of eleven journeys to the Arctic, I talked with several scientists to get a sense of the unfolding situation. None of them were hedging their bets about what was in store for the future. John Smol, the 2004 winner of Canada's top science award, the Gerhard Herzberg Gold Medal, was unequivocal about the perils that climate change posed for both the polar world and for people down south.

"We should be paying attention, but we're not," Smol told me. "Maybe it's because there are so few voters up there. Politicians have a difficult time appreciating that half of Canada's real estate is Arctic, and that two-thirds of its coastline is in the Arctic. On one level we have a responsibility to be stewards of this big piece of real estate. But even for selfish reasons, we should be concerned because the changes taking place up there are eventually going to catch up with us down here."

Mark Serreze was similarly cautionary in his forecast for the future. "The rest of the world will be in for a few surprises," he predicted. "What happens in the Arctic matters to the rest of the world. If we ignore what's going on, it's going to bite us down here, and it's going to bite us hard."

chapter one

NANUQ: IN THE TRACKS OF THE GREAT WANDERER

– Southern Beaufort Sea –

"He is almost an Inuk," the Eskimos say, discussing the ways of the polar bear. "He is the nearest to men."

—As told to Roger Buliard, an Oblate missionary who lived among the Inuit for fifteen years in the 1930s and 1940s

I AM STANDING ALONE on the Arctic sea ice watching the helicopter as it flies off toward a large body of open water. It is late in the evening in April, and the brilliant spring sunshine that was so blinding earlier in the day has now faded into a honey-colored glow of yellow and orange.

Alongside this big hole in the ice 60 miles (100 km) north of the coast, I see the ghostly silhouette of a polar bear being pursued by the helicopter. With remarkable speed, the animal lopes through the frozen landscape, appearing and then disappearing, playing a game of hide-and-seek among the jagged blocks of ice pushed up by the rising and falling tides.

Mesmerized by this enormous animal weaving its way through this polar desert, I am unaffected by the razor-sharp winds and the -4°Fahrenheit

(-20°C) cold that had numbed my face and hands earlier in the day. The unfolding scene has me so rapt that I begin to understand those Inuit stories of flying polar bears imbued with supernatural powers.

The bear, in all likelihood, does not know I'm here. But when it suddenly stops and turns in my direction, it is no longer so ghost-like. Emerging from the ice-crystallized haze suspended alongside the open water, it is now powerful, dangerous and, except for the helicopter hovering above, undisputed. I, on the other hand, am now cold, vulnerable, thoroughly unnerved and wondering what I am doing standing there alone on the frozen sea with the rifle still sitting unpacked at the bottom of a pile of survival gear.

I shouldn't have been there at all. A few days earlier I had gotten a call from Ian Stirling, the Canadian Wildlife Service scientist in pursuit of the bear that evening. He was sounding very apologetic and weary on the other end of a satellite phone. Two weeks into his survey of the southern Beaufort Sea polar bear population shared by Canada and Alaska, things were not going well.

"I'm hunkered down on Herschel Island just north of the Yukon coast." His voice quavered as it broke up during the transmission. "We've had an extraordinary amount of fog and whiteout conditions that have prevented us from flying most days. There's a lot of open water in the area and we're way behind schedule."

An uncomfortable pause followed.

"I'm starting to think that you coming up to join us may not be such a good idea. How much have you been counting on this trip?"

I wasn't quite sure what to say. This was supposed to be the first of eleven Arctic trips I had planned on for the next eighteen months and I was counting on it a great deal. Missing a chance to go out into the field with the world's leading polar bear scientist wouldn't necessarily be a disaster, but neither would it be a good way to start a project on how climate change is affecting the Arctic. Already an Arctic icon, polar bears have become the poster animal in the climate change debate. If anything about the meltdown of the Arctic mattered to the rest of the planet, it was the fate of this animal.

While I scrambled to collect my thoughts, Stirling began to reconsider. I could sense it in the way he was talking and going over his schedule. Stammering like the actor Jimmy Stewart, Stirling tends to think out loud and ask questions he eventually answers in a roundabout way. One minute into his ramble, he had my hopes up. The next, he dashed them. Finally, he relented, allowing me to win my case without having said a word.

"We'll be flying every chance we get in the next few days when we move out of here and if the weather cooperates, so in all likelihood there won't be anyone to pick you up when you get to Tuktoyaktuk," he said. "You'll have to make your way into town and get someone to direct you to the house we'll be renting. Just tell them that it's the Gruben house. They'll know where it is. There should be a key outside the door. You can use that to get yourself in."

Tuktoyaktuk is a small village of about nine hundred Inuvialuit of Alaskan and western Arctic origin who migrated to this area in the early twentieth century to trade with American whalers based at Herschel Island. Although it is remote, located just beyond the far end of the delta, Tuk, as locals know it, is not as impossible to get to as some Arctic communities. Jet service from the south goes to the larger town of Inuvik daily. From there, a small plane takes a handful of passengers 93 miles (150 km) north every day of the week if the weather permits.

The day I arrived in Inuvik, the plane to Tuk was running several hours late thanks to a bank of icy fog shrouding the coast. So rather than bide my time in an empty terminal building, I decided to try to hitch the 6-mile (10 km) ride into town to get what I suspected might be my last hot meal for a few days.

Inuvik is the last stop on the Dempster Highway, the only road in Canada that connects the Arctic to the south year-round, save for two or three weeks in spring and fall when breakup and freezing on the Mackenzie River prevents the ferry from crossing. Unlike most northern communities, Inuvik doesn't have a long or culturally significant history. The town rose up from the wilderness flats along the east channel of the Mackenzie Delta in the late 1950s when it appeared that erosion and flooding would eventually cause nearby Aklavik to slide into the delta. (To the delight of those who declined to be transplanted by the Canadian government, it never did.)

Compared to Juneau and Anchorage in Alaska or Pond Inlet and Pangnirtung in Arctic Canada, Inuvik may not be much to look at. There are no towering snowcapped mountain peaks, giant glaciers or undulating alpine landscapes hovering over the townsite. The spruce in the surrounding forests are small and skinny, and the mosquitoes can be ferocious on a hot summer day. But there are crisp, clear days in winter and spring when the Richardson Mountains heave into view and take the breath away.

The town also has good hotels, two or three half-decent restaurants and a first-rate recreation center. What makes it truly special is the mix of people who have settled there. Among its thirty-five hundred residents, there are Inuvialuit, Gwich'in, Métis and non-Aboriginal peoples. The Catholic priest is from Nigeria, most of the cab drivers are part of a small Muslim community intent on building a mosque and the medical director of the hospital is from South Africa. In summer, you can find many of them communing at the giant greenhouse the town constructed out of the old arena. It was built from the profits a former priest made after he secured the distribution rights to Coca-Cola in the region.

When I finally arrived in Inuvik, I found the bookstore closed, one of the hotels boarded up and Father Matthew Ihuoma, the Nigerian-born Catholic priest, away on church business. At the greenhouse, the doors were locked. Feeling a little sorry for myself, I stood alone on the snow-covered streets. Thankfully, my luck changed when I spotted Frank Pokiak, the chair of the Inuvialuit Game Council, walking down the road.

Frank, an Inuvialuit man in his late fifties, is the kind of person you never forget after you first meet him. Built like a football lineman, he has the bull-legged gait of a grizzly bear marching purposefully across the tundra. Not the kind of guy you'd want to rile. I'm sure he didn't recognize me from the two or three times we had met in the past but that didn't stop him from inviting me for a quick cup of coffee at a nearby hotel restaurant when I inquired about the possibility of visiting his beluga whale hunting camp that summer.

"Want some?" Frank said as he lifted a big plastic ice-cream pail onto the table.

"What have you got there?" I asked.

"Beluga *muktuk* in whale oil," he said as he opened the lid to give me a peek. "I brought it with me from Tuk so I could snack on it during our Game Council meetings here in Inuvik."

Swimming in the golden liquid were several chunks of chewy beluga whale blubber. By the fishy smell of it, I didn't need to get any closer to see if I wanted any. This batch was a little too fermented for my tastes. By the way the waitress homed in on me, then Frank and then me again, I could see that she found Frank's snacks equally unappealing.

"Just coffee," Frank said kindly. "And maybe a paper towel to clean up this small spill."

Like his three brothers, Boogy, Charles and James, Frank doesn't quite fit the stereotypical image of an Inuvialuit or Inuit elder. He and his wife, Nellie, have their own Internet site on Bebo, a social media network that allows friends to share their lives and explore entertainment opportunities. There, you'll find that Frank is a grandfather of three and father of five. He enjoys hunting and fishing and hockey. Polar bears don't scare him, but mice do. His favorite movie is *Legends of the Fall* and his favorite band is the Lost Bayou Ramblers. His grandkids, he tells me, are his hobby.

Frank and his brothers are well known to scientists such as Ian Stirling. For twenty-five years, they've assisted in a number of wildlife research projects in the western Arctic. Both James and Boogy worked with grizzly bear biologist Peter Clarkson, as did their brother in-law Billy Jacobson. For the past nine years, Frank has been collecting blood and tissue samples from beluga whales he harvests for Fisheries and Oceans scientist Lois Harwood, one of Stirling's former students.

Despite how savvy he is about science, Frank is first and foremost a hunter. Being a hunter, he worries about what will happen to the polar bears and beluga whales as the Arctic sea ice continues to melt. He's concerned that warming temperatures may be affecting the movement of caribou and other land mammals. But like many Inuvialuit, Inuit and Inupiat hunters, he's reluctant to buy into the more alarming climate change scenarios that some scientists have been forecasting for the Arctic world.

"The fact is our history is all about adjusting to new circumstances," he said. "Sure, it's disturbing to see birds and insects that you have no name for. And nobody thinks it's good that the ice is melting earlier than ever. But what can we do about it? We have to find a way of adjusting."

Few people in the western Arctic wield as much influence as Frank. He chairs the Inuvialuit Game Council that assigns community hunting areas and advises wildlife management boards and the federal and territorial governments on wildlife conservation issues and research projects. Scientists who don't impress members of the council almost always fail to get the permits they need to conduct research in this region.

This system of vetting has been a nightmare for some scientists working in the Arctic in recent years. After spending weeks, sometimes months, securing scarce resources for a research project, more than one scientist has found themselves unable to convince the Game Council, or other similar advisory bodies in the Arctic, of the worthiness of their project. Sometimes, it's a legitimate concern about immobilizing animals with tranquilizers if the person doesn't seem qualified or sufficiently experienced. In other cases, the opposition arises out of concern the research will lead to a ban or reduction on hunting polar bears, beluga, Peary caribou and other Arctic animals.

No animal more than the polar bear is a lightning rod for this kind of skeptical scrutiny. Not only does the polar bear provide the Inuit with food and clothing—it puts money in the pockets of many Inuit guides. American and European sports hunters will pay as much as $35,000 for a guided polar bear hunt. So any research that warns of declines in polar bear populations is often seen as an attack on the Inuit way of life.

Ironically, that traditional way of life very much parallels the polar bears they harvest. Like the polar bear, an Inuit hunter has to be creative in developing harvesting strategies to take advantage of the different situations and habitats he encounters during a year. He also has to be resourceful not only because seals and other marine mammals are relatively scarce over the Arctic range, but because they are so good at escaping. To have survived in this polar world, the Inuit had to be nomadic and forever investigative, just like the animal they hunted. That's why the Greenland Inuit call the polar bear *pisugtooq*, the great wanderer.

Given their relationship with the polar bear, it is not difficult to understand why the animal is such an important character in Inuit folklore. One of the more popular tales tells of a ten-legged bear that was able to dupe hunters into mistaking its abnormal anatomy for a group of humans walking in the distance. One after another, the bravest of hunters who went out to investigate the strange sight disappeared. In the end, an elder, still wise but little respected, tracked down the ten-legged animal to its hiding place and tricked it into an ice passage that had no exit. Unable to find its way out, the bear retreated to the opening, where the old man was waiting with a spear. The killing of the animal restored the man's reputation among his people.

Although stories such as this one reflect respect for the power and intelligence of the polar bear, they also convey the notion that the killing of a polar bear is a symbol of a man's hunting prowess. This is proving to be increasingly problematic in a world where an orphan cub such as Knut, the bear that saved the flagging fortunes of a German zoo, became an international celebrity overnight. Southerners may be sympathetic to Inuit efforts to preserve their culture, but as their opposition to trapping and sealing showed, animal welfare tends to win their hearts.

The world's love affair with the polar bear is a relatively recent phenomenon. In the late nineteenth century, European and American whalers were ruthless in their indiscriminate killing of the animals. One whaling captain boasted that he shot thirty-five polar bears in the Amundsen Gulf of the southern Beaufort Sea, presumably for lack of anything better to do. By modern standards, the callousness of many of these kills is incomprehensible. An anonymous surgeon aboard the *Hercules* in 1831 described how, for sport, he and others chased down a female and her two cubs in the open water:

> I fired at the old one but missed. They all swam close together; the mother appeared to caress her cubs with her nose. When we approached within 30 yards of them, the mother turned and charged open-mouthed. Candy, our harpooner, pushed the lance into her neck. She took the lance in her teeth and dived and came up roaring among our oars . . . (Finally) Milford finished her off. She dropped her head into the water, her last look being directed toward the cubs who were blowing out and swimming to sea. We shot one of the cubs with a bullet. The other kept swimming around its brother. Milford threw a noose over the other's head and drew him up to the boat's bow, where he hung roaring and biting the boat's stem. He was strangled before reaching the ship.

Thomas Macklin, a surgeon aboard another whaling vessel, was even more cold-blooded when his ship encountered a bear standing on the sea ice as they sailed toward it:

> He must have been hungry, poor fellow, for instead of trying to run away, he stood his ground, intending, no doubt, to make a meal of the strange black bird (the ship) with huge white wings; but the flame shot forth from her head and he was brought to his knees in supplication for mercy. But his life's blood was necessary to expiate the crime of arrogance and intrepidity in daring to approach so mighty a bird; again flame shot forth and the messenger of death demanded of him his life. Such is the punishment which awaits prideful arrogance.

In the short time they plied the waters of the Arctic, the whalers may have been just as effective in depleting the polar bear population in the Arctic as they were in driving bowhead, beluga and narwhal whales to the point of extinction. No one knows for sure. Up until 1935, there were no laws to protect polar bears in North America nor any means of monitoring their numbers. Canada's decision to limit the hunt from May through October that year was one of the first attempts at conservation. In spite of this and other similar measures taken by the United States, Norway, Greenland and the U.S.S.R. in the ensuing years, the numbers continued to dwindle into the 1960s.

The decline of polar bears in the twentieth century is linked, in large part, to technology. First the rifle and then the snowmobile in the 1960s fundamentally changed the relationship between the Inuk and the polar bear, reducing the classic confrontation between man and beast to an arcade shooting game. In 1948, only 148 polar bears were killed in Canada. By 1960, that number rose to 509. By 1967, when the snowmobile overtook the dogsled as the primary means of Arctic transportation, more than 700 bears were killed. The trend, which was worldwide, was exacerbated when sports hunters and trappers stepped into the game with aircraft and preset guns to help them take their prey.

By the middle of the 1960s, scientists such as Dick Harington, the paleontologist who, remarkably, started off as polar bear scientist with the Canadian Wildlife Service, estimated there might be ten thousand or so bears left in the world—a little less than half of what there is today. The alarm led to the first circumpolar meeting in Fairbanks, Alaska, in 1965, at which a resolution was passed protecting females and cubs. A further step, taken in 1968, brought the same nations—the United States, Canada, Denmark, Norway and the Soviet Union—together under the auspices of the International Union for the Conservation of Nature for the first of a continuing series of meetings that took place every three to five years. At those meetings, the circumpolar nations agreed to pool their resources and research efforts to ensure a future for the species.

Ian Stirling was among the first scientists to focus on the polar bear population in the southern Beaufort Sea in a meaningful way. Although he is too modest to take credit, his research helped convince Inuvialuit leaders and government managers of the need to stop hunting female bears that were accompanied by cubs. His research also emphasized the importance of not hunting in the fall, when pregnant females are searching for dens. The number of bears each community is allowed to harvest each year rely heavily on the population estimates that he and his colleagues made from their surveys.

One of Stirling's biggest contributions arose from the long-term tracking of the animals he and his team tagged and captured in the Canadian Beaufort Sea. For a time, it was thought that the bears living on the Alaskan and Canadian sides of the international border were distinct. So for decades, both the Canadian and U.S. governments managed the polar bears of the southern Beaufort Sea in their own way.

But by the mid-1980s, Stirling and Steve Amstrup from the U.S. Geological Survey recognized the potential pitfalls of this dual form of management. After crunching their numbers from years of mark and capture efforts and from the radio collar tracking that Amstrup successfully used to follow the movements of animals in the region, they discovered that the two populations were really one. Many of the estimated 1,800 to 2,000 animals in the region were moving back and forth across the border on a regular basis. In one notable example that Amstrup documented years later, a female polar bear from northern Alaska walked westward for four months before reaching northern Greenland. To the amazement of many, she came within 2 degrees of the North Pole.

The movement of polar bears between Canada and the United States was particularly worrisome for both the Inuvialuit and the Inupiat and for government managers on both sides of the border because conservation measures at this time were somewhat archaic. While Inuvialuit hunters were bound by a legally enforced quota system that dictated how many animals each community could harvest, for example, the vagaries of the U.S. Marine Mammal Protection Act made it possible for the Inupiat in Alaska to hunt as many animals as they wanted so long as it was for subsistence purposes. Another anomaly in the U.S. act allowed them to harvest females in dens. The rules were so bizarre that the U.S. government could do nothing until the polar bear population was considered to be "depleted."

Fortunately, the potential for overharvesting polar bears in Alaska at the time was minimal. But it was a serious concern for Inuvialuit leaders such as Andy Carpenter, Alex Aviugana and Nelson Green of Canada. Like their Alaskan partners, they did not wish to see the population overharvested. Nor did they want to risk losing the lucrative business that American sports hunters brought to the Canadian North. In order for those hunters to legally bring back their polar bear hides, the U.S. government insisted that the country in which the polar bear was shot had to have solid scientific data showing that the population was being managed in a sustainable way.

After recognizing the problems on both sides of the border, representatives from the Inuvialuit and the Inupiat began to meet to decide how to manage and conserve the southern Beaufort population as a single unit. After just two years, the Inuvialuit Game Council of Canada and Alaska's North Slope Borough met in Inuvik in 1988 to finalize a gentlemen's agreement to manage the population jointly.

Unfortunately, the deal, the first of its kind, did not get off to a great start. Not long after the ink was dry, a young hunter from Alaska shot a female, leaving her two cubs orphaned. Too young to be on their own, they were placed in a zoo. Worse yet, the lost animals did not count against the quota of bears the community was allowed to harvest. Nothing legal could be done to punish the hunter, but the Wildlife Management Department of the North Slope Borough and its partners on the Canadian side of the border made it clear that actions like this would not be tolerated in the future. So when a hunter from Inuvik shot and killed a female and her single cub near her den site in 1994, charges were laid. The Inuvik Hunters and Trappers Association also suspended his polar bear hunting privileges for five years.

Frank Pokiak is proud of the fact that the agreement has worked so well since then. But like many hunters in the region, he is not optimistic about what the future holds for polar bears in the Beaufort Sea. In recent years, hunters from Tuktoyaktuk and other western Arctic communities have not harvested all the animals they are entitled to take.

"The ice is changing," he told me. "It melts earlier in the spring and comes later in the fall. Hunters have to go farther and farther away to find the bears. Some are giving up. With the high price of gas, it's getting too expensive. It's also dangerous because the ice is getting too soft in springtime."

What worries Pokiak even more is the day when scientists such as Stirling and Amstrup will bring data that suggests that what few animals they are finding and harvesting is not sustainable in a declining population. How the Inuvialuit and Inupiat respond to that scenario, he admits, will be the toughest test of their co-management agreement.

The Arctic covers an enormous amount of landscape, but hot news travels fast. Oftentimes, pilots deliver tantalizing bits of news and gossip long before it's shared over the radio, Internet or telephone. On the short trip to Tuk, the pilot was buzzing with reports about an American hunter who shot a strange-looking animal on the sea ice just west of Banks Island. "I heard that it looked like a polar bear and a grizzly bear mixed into one," the pilot told me. "The wildlife officer was apparently so confused that he seized the animal." I dismissed the story as another of the tall tales you often hear in the Arctic from people who haven't been there long, but it proved to be true. An American hunter shot a bear that turned out to be polar bear/grizzly cross, the first one ever found in the wild.

The pilot also told me of a grizzly bear denning north of the airstrip at Sachs Harbour on Banks Island—hundreds of miles north of where the brown bears are normally found. This anecdote had an intriguing ring of truth. The number of sightings of grizzly bears on the sea ice has been rising steadily since biologist Mitch Taylor spotted one several years ago in the farthest reaches of the High Arctic where no grizzly had ever been seen before.

Taylor was flying over Viscount Melville Sound at the time doing a routine survey of polar bears when he spotted the bloody remains of several seal pups on the sea ice. This in itself is not unusual. Typically, a polar bear will leave everything but the fat of a seal behind. A big healthy bear will sometimes take a single bite and move on. Nearby, however, Taylor also found the remains of a young polar bear that had evidently been killed at the same time.

This was also not a big surprise. Sometimes polar bears will engage in vicious battles that leave one of them dead or badly injured. Some have also been known to engage in cannibalism, especially when they're in bad shape and starving. Determined to find out what happened in this case, Taylor and the helicopter pilot followed the killer's tracks until they spotted something dark moving in the distance. Thinking the light was playing tricks on them, Taylor signaled the pilot to fly in for a closer look. Instead of running away as most polar bears do when they hear the sound of an engine, this animal abruptly turned and ran defiantly toward the aircraft. That's when Taylor realized it was a grizzly bear.

Since then, more and more grizzly bears have been spotted on Banks and Victoria islands and as far north as Melville Island north of Viscount Melville Sound. What the animals are doing in the Kingdom of Nanuk, the Great Wanderer, no one knows for sure. But looking out at the white expanse of the Beaufort Sea as we began descending into Tuk, I wondered whether the young grizzly bear Taylor had caught that day was the same one denning on Banks Island.

Tuktoyaktuk is a derivative of *tuktu*, the Inuvialuit word for caribou. Legend has it that an old woman was sitting on the coast one day when she spotted a group of caribou wading across the tidal pools in front of her. Using her shamanistic powers, she petrified the animals so that she could gaze on this bountiful site forever. Today some people in Tuk insist that in summer, you can see the shapes of the animals in the reefs that rise up from the town's narrow harbor during low tide.

Tuktoyaktuk's ties to its cultural past are fractured ones. More than any other community in the Arctic, the people there have been hardened and made savvy by their long and often difficult association with non-Aboriginals from the south. First, it was the whalers who introduced alcohol and diseases that the Inuvialuit were not immune to. Then it was the oil and gas workers who invaded the community in the 1980s with more alcohol, drugs and empty promises of prosperity. Most people are still waiting for the good times to come.

Not that the wait has been without drama. Several years ago, Molson, the beer giant, staged a beach party in Tuk on the Labor Day weekend. The contest for tickets brought hundreds of winners from all over the continent to hear several rock bands, including Hole, Metallica and Moist. People in town still talk about grunge diva Courtney Love staggering off the plane barely able to sign autographs for contest winners and then stumbling on stage the next night making fun of Tuktoyaktuk's name by replacing the first and last syllables so that it rhymed with a profanity.

Well aware of all that has transpired in Tuk over the years, I was still expecting to observe something traditional on the weekend I arrived. It was the Beluga Jamboree, when people in the community come together to celebrate their whaling history. But when I strolled down toward the harbor that evening, there were more iPods and super-charged snowmobiles than drums and harpoons. I felt a little like Danish ethnologist Knud Rasmussen, who, finding an Inuit tent camp in the most remote corner of the barrenlands in 1922, was disappointed to discover its inhabitants listening to the sound of Caruso's mighty voice ringing out from a gramophone. He was, he noted at the time, a hundred years too late.

"Brutal is the only way I can describe the field season so far," Ian Stirling told me when we finally met that night. "We're not finding nearly as many bears as we would have expected. And the weather hasn't been cooperating either. In the six hours we flew today, we caught just two bears. Dennis [Andriashek, Stirling's longtime wildlife technician] hasn't had much luck either. He's been sitting in the fog for days at Sachs Harbour waiting for the weather to clear. When it finally did yesterday, the Royal Canadian Mounted Police (RCMP) put the pilot and the helicopter on hold just in case they needed it to find a hunter who had gone missing. Turns out that the hunter was at home sleeping in bed.

"This," he said wearily, "is rapidly turning into one of the worst years we've had."

We were in a house that was typical of the small bungalows found across the Arctic. Four of us shared the rented place: me, Stirling, Evan Richardson and Andrew Derocher, the Canadian scientist who chairs the polar bear specialist group for the International Union for Conservation of Nature (IUCN). The main room of the house was long and narrow and divided into kitchen and living room. A hallway led to three tiny bedrooms and a bathroom that was serviced with trucked-in water stored in a big tank. The only thing that distinguished it from any other place I had been to in the North was the enormous pair of head-banging speakers blocking the view through the front picture window. Dwarfed by these behemoths, Stirling was sitting on an oversized couch. When he saw me staring at the speakers, he assured me that they belonged to the owner, not him.

Since it had been some time since I had last seen Stirling, I thought he might have aged. But apart from a few more gray hairs in his beard and a thinning hairline that made him look more professorial than he already did, he seemed to be as fit as ever and far too young to be the emeritus scientist he was about to come.

Few individuals in the field of Arctic science cast a longer shadow than Stirling, whose career, ironically, got its start at the other end of the polar world forty years ago because jobs in the Arctic were scarce. Since then, his research on polar bears, seals, walrus and other marine mammals has earned him the kind of international respect that very few environmental scientists command. In addition to the nearly two hundred peer-reviewed papers he's published in dozens of scientific journals, he's written three popular books. When the U.S. government was considering putting the polar bear on the threatened list, Stirling was asked to be the lead author of one of the seven reports commissioned and co-author of two others that eventually went to the U.S. Secretary of the Interior, Dirk Kempthorne, for consideration.

Andrew Derocher refers to his mentor as "Mr. Polar Bear." What sets Stirling apart from the rest of the scientific crowd, he told me that evening, "is that he is much more than just a number-cruncher trying to figure how many bears can be harvested in a given year. Ian," he said, "is always looking at the bigger picture in the ecosystem to see how bears, seals and other marine mammals fit in."

As much as Stirling enjoys his job, there are many downsides to flying in cold, windy or foggy weather. Over the years, Stirling's crew has been involved in two helicopter crashes, two fixed-wing crashes, two helicopter engine failures and more narrow escapes than he'd care to talk about. Fortunately, no one in his group has been hurt. But Malcolm Ramsay, a former student who went on to distinguish himself in the field of polar bear science, died in a helicopter crash in Lancaster Sound while coming home from his own project one spring.

Stirling also had a close call when he was testing an antidote Ramsay thought might help tranquilized bears recover more quickly from the drug scientists once used to immobilize them. With the two of them at the time was Canadian Wildlife Service veterinarian Eric Broughton and veteran helicopter pilot Steve Miller. Not knowing how the tranquilized bear they captured would respond to the antidote, Stirling watched cautiously from a distance after Broughton injected a small amount of the drug into a vein beneath the animal's tongue. After several minutes, they prodded the bear to see if it was recovering. There was no response.

Stirling then knelt down and pulled on the bear's tongue, as he would typically do to see how deeply tranquilized the animal might be. Again, there was no response. A few minutes later, he tried again, assuming that nothing would happen. But this time he made the mistake of reaching into the mouth from the front instead of the side, as he always does to avoid the possibility of the bear's sharp front teeth clamping down on him. Just as he grabbed the tongue, the jaws snapped tight, piercing the sides of his thumb and forefinger and just missing the bone. The antidote was working. As the bear's eyes became more alert and focused, Stirling knew he was in serious trouble. Eric Broughton tried to pry the bear's mouth open but was unsuccessful. Only when he slapped the bear on the nose did the bear loosen its grip long enough for Stirling to pull his hand out.

Close calls like this one haven't deterred Stirling from trying to do everything and just about anything to help a bear that has had difficulty recovering from those early experimental drugs. One notable example of this diligence occurred on Banks Island more than twenty-five years ago when he darted a thin adult female bear a little too heavily. When Stirling discovered that the animal had stopped breathing, he did what some might consider unthinkable. He started applying artificial respiration, pausing to check for light breaths for any sign of recovery.

Because the wind was blowing so hard that day, Stirling had trouble detecting even the slightest movement in the animal's rib cage. So he placed his ear against the bear's nose and cupped his hand over the top to make a better seal for his breath to go through.

Several minutes after the bear was obviously breathing on its own, Stirling looked up and noticed that his colleague had backed up several feet, shifting back and forth nervously from foot to foot. Curious about this strange behavior, Stirling was about to ask him what was up when his friend finally blurted out: "You know, if you keep doing that, you're not going to live very long."

Fortunately for the bears and the scientists, a newer, safer drug has come onto the scene. Not only does it immobilize the bear more quickly and allow for a speedier recovery, it makes it easier for scientists to tell when it is safe to approach a tranquilized animal. In the final stages of the induction, bears tend to hold their heads up before quietly lying down.

After thirty-seven years of this sort of drama, Stirling has come to accept the fact that not all goes according to plan when you're trying to figure out how many animals there are in an extremely remote Arctic region such as the Beaufort. If fog, dangerous ice conditions or stormy weather don't set you back a week or so, as happened to him and his crew this spring, then mechanical trouble will almost certainly take the wind out of your sails. He's also experienced a year like this in the western Arctic when bears seem to be few and far between.

"The winter of 1974–75 was an especially cold one that created ice conditions that were too unstable for ringed seals—the polar bear's primary food source—to successfully build birthing lairs," he told me that night. "It was a real puzzle at the time and a scary one because initially we couldn't figure out where all the bears had gone."

It was cold and clear when we finally set off the next morning, but a bank of fog that loomed large on the horizon had Stirling wondering out loud whether we would be able to put in a full day. A week without snow or hard wind, he added, was also going to make it that much harder to find tracks.

Strange as it might seem, it is much easier finding a polar bear by following its tracks than it is scanning the horizon for any sign of this animal that can, at one end of the extreme, weigh as much as 1,800 pounds (800 kg). The polar bear's paw, which is as long as 1 foot (30 cm) across, is not only big and easily spotted at 300 feet (90 m), it offers signs of what the polar bear is up to. The tracks of an animal moving in a straight line before suddenly turning in another direction and then stopping often signals that a bear has picked up the scent of seal. The wide tracks of a fat bear in spring will differ from that of a skinny one in fall, when many of the bears have gone for some time without food. And snow that has piled up on one side of a footprint might mean that the track is not all that fresh.

The bear's paw is a remarkable thing. It helps the animal distribute its weight when walking across thin ice. And papillae, those soft bumps on the black footpad of a bear, keep the animal from slipping, as do the claws, which can be as long as 2 inches (5 cm). Although the polar bear is well equipped for walking on slippery ice, it prefers to move along patches of snow, often stopping to sniff out the chubby ring seal pups still in their lairs at this time of year.

We were more than two hours into the search when Stirling finally picked up tracks that showed promise. But any hope of catching the animal faded quickly when the footprints ended abruptly at the edge of a large body of open water.

"A lead like this is a magnet for bears because this is where you tend to get a lot of seals hauling out," explained Stirling, sitting in the front seat of the helicopter. "So there's no surprise that our guy has gone for a swim. The challenge is figuring out where he resurfaced. With the calm, clear weather we've been having lately, it's almost impossible to determine which of these tracks we're seeing on the other side of the lead are fresh and which belong to our bear."

In their never-ending search for food, polar bears travel long distances across ice, often jumping into open water and swimming for several hours to get to where they want to go. Refueling at Baillie Island, a small, uninhabited rise of land that was once home to some families from Tuk, Stirling told me how he once watched a polar bear hunting for several days in Radstock Bay in the High Arctic when it suddenly changed direction, plunged into the water and then disappeared.

"The next day I got a radio message from a colleague who, by chance, spotted the same marked animal along the north coast of Somerset Island, a 100-kilometer [60 mile] swim away."

The fact that polar bears move such long distances over extremely remote areas makes it expensive and difficult to estimate how many animals there are in the thirteen regions of North America where the polar bear is found. As things stand now, Canada and the United States have only good long-term data on two of these populations—the one in southern Beaufort Sea and the one that makes its living in Hudson Bay. What is known about the rest is based on fragmented studies and educated guesswork.

The big picture is not a particularly promising one. Five of the thirteen populations appear to be on the decline. At least one—the Baffin Bay population—is the victim of severe overhunting on the Greenland side of the polar bear's range. Of the three that are increasing, the Viscount Melville and M'Clintock Channel populations are on the rebound only because of the Canadian Inuit's voluntary decision to severely reduce their harvest.

It may well be that the deteriorating ice conditions that come with climate change are now beginning to take their toll on bears and seals in this part of the world. If that turns out to be the case—as more and more scientific data suggests—then those in the very small minority who insist that climate change is no threat to polar bears are plainly wrong. It also suggests that there needs to be a fundamental rethinking about how polar bears are hunted and managed and how much greenhouse gas we're pumping into the atmosphere.

"We've documented a 22 percent decline in the western Hudson Bay population between 1987 and 2004," noted Stirling.

"The animals that we see there now are younger and thinner than the typical bear you'd see twenty or thirty years ago. Martyn Obbard from the Ministry of Natural Resources in Ontario has observed a similar situation in southern Hudson Bay. The reason is pretty simple. Bears pile on the fat they need to make it through the year by catching seals on the ice. With the ice melting two or three weeks sooner in spring, as has been happening in western and southern Hudson Bay, the animals are spending more time on land and getting less opportunity to put on the reserves they need to successfully reproduce and to make it through the year. It's a double-edged sword. Less time feeding also means more time burning up stored fat."

In the politically charged world of polar bear science, Stirling is reluctant to categorically conclude that the 15 percent decline that he and his Alaskan colleagues are observing in the Beaufort Sea is similarly linked to climate change. There simply isn't a smoking gun. But adding up all the telltale signs, he points out—the recent drowning of several bears off the Alaska coast, the fact that bears in the Beaufort are not as fat as they once were, the apparent increase in cannibalism, the trend for more bears on the Alaska side of the border to give birth on land rather than on the ice, the warmer weather and changing ice conditions—suggests that climate change could now be having an impact on marine mammals in the western Arctic of Canada and the United States.

Stirling is hardly alone in waving the red flag. When he and a select group of scientists finally finished their reports for the U.S. Geological Survey, they concluded that two-thirds of the world's polar bears, including all those in Alaska and most of Canada's western Arctic, will be gone by 2050. The only ones remaining, they added, would be those inhabiting the High Arctic regions of Canada and western Greenland.

"You see that down there," said Stirling, pointing to a spot where a bear had evidently spent a great deal of time digging down into the ice. "That's what we call scratching. It's normal for bears to dig into the snow.

But a bear has to be really desperate to spend all that energy scratching through the ice trying to get into a seal lair. We saw ten examples of that last year and three so far this spring. In most cases, it was a waste of time and energy for the bear."

Now that the polar bear has become the poster animal in the climate change debate around the world, one might have expected that Stirling and his colleagues would have no trouble getting the funding to answer some of the questions that need to be addressed as the ice quickly recedes in the Arctic and as energy and mineral exploration escalates. But Stirling says he still feels like an "organ grinder with a monkey on his back begging for money. . . . As much talk as there has been about climate change and polar bears," he told me, "there has been no corresponding increase in government funding for polar bear research." The Americans who co-manage the population, he points out, are funding much of what he is now doing in the Beaufort Sea.

The funding situation is especially crucial because climate change is just one of a number of things threatening the future of polar bears. Mounting evidence suggests that toxic chemicals such as PCBs, PBDEs, DDT, dioxin and dieldrin and heavy metals such as mercury are contaminating the polar bears' food supply. Hunting pressure on bears is also intensifying, as are conflicts between bears and humans in Arctic zones that are undergoing energy development.

None of this has silenced the contrarians who continue to insist there is no reason to be concerned about the future. The claims—some of them supported by petroleum interests and big-game sport hunters, some by the Inuit themselves—are coming from all directions. On the one hand, these critics insist that dips in population are symptomatic of natural variability—not greenhouse gas emissions. On the other, they argue that polar bears will adapt to a life on land eating berries, salmon caribou and snow goose eggs. Then there are those who point to a recent study showing the Davis Strait population in the eastern Arctic increasing. They say the numbers prove that scientists such as Stirling are wrong.

Stirling bristled at the mention of this.

"The fact is we [the IUCN polar bear specialist group] have been saying all along that the Davis Strait population was likely on the rise," said Stirling. "It's also possible that other populations might increase in the short term as the Arctic gets warmer and possibly more biologically productive in some areas. It's what's going to happen in three or four generations—forty-five years or longer—that really worries us.

"Superficially, polar bears may appear to be secure in some places. But everything we've learned about them over the past three decades suggests they will not fare well at all in a world with little or no ice for a good part of the year. Polar bears live almost entirely on a diet of seals. To suggest that they can suddenly switch and find enough food eating berries or salmon or hunting down caribou, as some people say, is fanciful. We've observed what bears do on land during summer and fall in western Hudson Bay when the ice has melted. It's pretty clear that they aren't doing much but sleeping, fasting and waiting for the ice to form. The really hungry ones end up going into the town of Churchill to look for food."

Stirling doubts that the range of the polar bear will significantly shift north as things warm up.

"Right now the area north of where we normally find bears is a comparatively sterile environment. It remains to be seen whether it will become significantly more productive as the ice recedes and things warm up. My hunch is that there may be a small refugium for a limited number of bears in the future, but nothing to make up for the collapse we'll see in southern latitudes of the sub-Arctic and Arctic world."

Stirling understands why some Inuit and Inuvialuit hunters have a difficult time accepting his and his colleagues' findings. It not only means that they will have fewer polar bears to hunt themselves, but also fewer bears to give up to the lucrative sport hunting industry. He also accepts the Inuit contention that they are seeing more bears in some places than they have before.

But those anecdotal observations don't take into account the fact that polar bears often have a home range of more than 77,220 square miles (200,000 sq. km) or that the bears they are seeing are the result of sea ice changes or large-scale shifts in the distribution and abundance of prey.

"Some well-intentioned people have suggested that we start feeding the bears to help them through the ice-free season," said Stirling, pointing out that there is no easy way of saving polar bears. "But polar bears are large animals. The average bear requires about forty-five to fifty ringed seals a year to survive. Putting it crudely, twenty thousand polar bears would require nine hundred thousand ringed seals or the equivalent. That's a lot of seal.

"The bottom line is that hunting quotas need to be conservatively set to reflect not only the bear populations, but all of the other changes that are occurring in the Arctic world. And if the world wants to save this iconic species, humans are going to have to reduce greenhouse gases very quickly and very substantially."

Come early evening, we were still searching in vain for a polar bear to emerge at the end of the tracks. A foggy haze softened the blinding light of the ever-circling Arctic sun. Watching the muted shadows along the wind-carved ridges of ice below, I was tricked more than once into thinking I was seeing a bear. Looking out at this silent, still and alien world, I couldn't help wonder how a polar bear—even with its white camouflaging fur, its elongated head and body and its short, curved claws and large feet designed for efficient walking on ice or swimming in water—could thrive. In the hours we flew late that afternoon and into evening, there wasn't a single living thing to be seen.

And then, just as it seemed like we might have to go back to camp without having caught a single bear, Stirling picked up the fresh tracks of an animal, which, like the bear we followed earlier, were heading toward open water.

Wasting no time or words, he turned to me and asked, as he had at the beginning of the flight, whether I was comfortable being dropped off onto the ice while he tranquilized the animal from the air.

"Like I said before, I can't afford your weight and all the gear we have on board hampering the pilot's ability to maneuver quickly and safely over the animal. And remember, don't forget the gun. Pull it out of its case and have it ready. The chances of you needing to use it are remote, but you never know."

Stirling may not be a big man, but there is something intimidating about a man who has spent so much of his adult life tracking down and handling North America's largest and most fearsome animal. So when the helicopter set us down on the sea ice that evening, I jumped out and lowered my head to get below the rotor blade. Then I circled around front to get to the small cargo door. Methodically, I pulled out the survival gear, the gun and a barrel of fuel to make the chopper as light as possible for Stirling while he was up in the air aiming the tranquilizer gun at the bear. I was determined not to screw up.

All went according to plan until I tried to remove the fuel drum. To my dismay, it was too tall to slide directly out of the storage chamber. Nor could I lean it over sideways because the fuel pump was blocking the way. Stirling, I figured, must have been in a fury watching the bear running away while I fumbled about. Realizing how every second counted, I finally jerked and tugged every which way before the drum somehow slid out and fell onto the ice. Then with my back to the helicopter, I pulled my hood up over my head to avoid the blast of snow that ensued when Stirling and the pilot lifted off.

I was completely out of breath when I turned and saw the polar bear running toward me. In a panic I went for the gun that I had forgotten to unpack. Only when I looked up again did I realize that the pilot was simply maneuvering the tranquilized bear away from the open water so it wouldn't jump in and drown. And then just like that, the bear running toward me lay down, raising its head just once before it dropped back down on the ice. The high drama that seemed like an hour to me took all of two minutes.

Stirling was patting down the animal, feeling for fat on its spine and hipbones, when I arrived on the scene.

"He's a solid two out of five," he said, looking the animal over. "The only fat reserve he has is on his butt. It's another sign that these animals are having a hard time finding seals. Normally at this time of year, we'd expect to see three. Question is, are there fewer seals or are they just harder to get at in these difficult ice conditions. Either way it's hard on the bears."

Back at the house in Tuktoyaktuk, Stirling was taking notes and reviewing the week's work.

"Forty hours of flying and just eight bears," he said as he looked up. "That's a pretty big investment for such a small return."

Looking over his notes from the previous year, Andrew Derocher put it all into perspective. "Last year by this time, we had 258 bears," he said. "This year, we have just 58."

The possibility that the bears had moved north to find the ringed seals was a possibility that Stirling wasn't counting out, and he suggested that Derocher might want to shift his search efforts there when he takes over later in the week.

Derocher agreed, but he was not optimistic.

"Maybe the bears have moved north or dispersed in other directions," he said. "If that's the case, then they're being forced to walk another 200 or 300 kilometers [120–180 miles] to compete for meals that other bears in the area are already looking for. You have to wonder what's going to happen down the road as the ice continues to retreat. As that margin of favorable ice keeps moving north in the coming years, it will only be a matter of time before these bears run out of room."

chapter two

THE LOST WORLD

– Brintnell Glacier, Northwest Territories –

The glaciers creep
Like snakes that watch their prey, from their far fountains
Slow rolling on; there, many a precipice
Frost and the Sun in scorn of mortal power
Have piled: dome, pyramid, and pinnacle,
A city of death, distinct with many a tower
And wall impregnable of beaming ice.

—Percy Bysshe Shelley, "Mont Blanc," 1816

In the summer of 1955, a small group of American rock climbers were dropped off at Glacier Lake near the edge of a huge ice sheet straddling the Continental Divide along the Yukon–Northwest Territories border. They were hoping to find some virgin mountains to scale. When the group finally found the cluster of jagged peaks and sheer rock walls they were searching for, they were haunted by what they saw rising up before them. It looked like the craggy spires of Yosemite, which Ansel Adams made so famous during his mountain photography expeditions in the 1920s, had been transplanted 1,860 miles (3,000 km) north.

Expedition leader Arnold Wexler had evidently underestimated the difficulty of what he had in mind that summer. Just getting to the site had been an expedition in itself. First, they had to drive 634 extremely rough miles (1,020 km) of Alaska Highway before flying in by bush plane to Glacier Lake. The hike from there was another two days of bushwhacking through thick brush, crossing icy streams and picking their way across slippery, snow-covered rock falls. When they finally got over the shocking head-on view of the 9,000-foot-high (2,740 m high) fins of wind and ice-polished granite that stand tall and angular facing one another in a half-circle, the adjoining mountains suddenly looked small and terribly ordinary. Turning toward his partners, Wexler declared most of the peaks "unclimbable."

Still largely untrodden, the Cirque of the Unclimbables is legendary in the world of mountaineering. Royal Robbins, the great American climber who went on to found a successful outdoor clothing company, managed, with some cunning, ingenuity and daring, to find a route up Mount Proboscis, the tower of granite that looks like Yosemite's Half Dome. Those few who have successfully crawled up the Cirque's curious-looking Lotus Flower Tower have described it as one of the greatest climbs in the world. For the elite rock climber, the Great Canadian Knife is one of a number of peaks in the area that still offers fresh lines for rock climbers.

Mike Demuth won't deny that he had the Cirque in mind when he got the idea of skiing across this much smaller icefield on the Yukon–Northwest Territories border. Beginning on the Bologna Creek side of Mount Mulhulland, the 9,823-foot-high (2,994 m high) peak that splits the east arm of the icefield in two, his plan was to ski over the icy plateau for three or four days before dropping down into the Fairy Meadows, the enchanting alpine valley that lies at the foot of the Cirque. But the glaciologist from the Geological Survey of Canada had more than adventure in mind when he put together a plan. His main goal, he explained when he invited me to come along, was to find out how quickly this icefield has been retreating and to determine what effect it is having on weather patterns and downstream river flow on the South Nahanni River in Nahanni National Park.

Icefields are the accumulation of hundreds, thousands and often tens of thousands of years of snowfall that compresses into ice over a relatively flat area of land in mountainous territory. As these icefields expand, they eventually spill over the edge of a plateau and become glaciers. These rivers of ice eventually melt into streams and rivers. Icefields such as those in the Rocky Mountains of Alberta and British Columbia and in the Wrangell–St. Elias and Kluane Range in the Yukon and Alaska are massive. Some can be hundreds of square miles in size and more than half a mile (1 km) thick. They are so cold and so big and moisture-laden that they can make their own weather.

"As far as I'm aware, this is the first time anyone has tried to do this trip on skis," Demuth told me. "There will be four of us if you come along: you, Steve Bertollo, who has been assisting me in the field for many years, and Matt Beedle, a ski bum from Juneau, Alaska, who is doing his Ph.D. in glaciology. We're going to be spending two weeks on glaciers in the Rockies before driving up the Alaska Highway to get to Tungsten on the Yukon–Northwest Territories border. It's about a half-hour flight into the icefield from there. You can come with the helicopter that will pick us up from the other side of the mountain divide at Fort Simpson. All you need to do is get to Fort Simpson to catch a ride in with the pilot."

During the thirty years that I'd spent pursuing outdoor adventures, I had canoed and kayaked down dozens of Arctic rivers and skied and climbed in some of the remotest regions of the barren lands, the Arctic islands, the Yukon, Alaska and Siberia. But the art of skiing on glaciers was still a challenge that I had yet to master. I had done only two trips at high elevations in the St. Elias Mountains and just one other in the Columbia Icefields of the Rockies. While I enjoyed each one of them, the second trip to the Kluane Icefields made me think twice about doing it again. Three-quarters into that week-long expedition, an unexpected wild spring snowstorm trapped the four of us on the pass that connects the Lowell Glacier to the Kaskawalsh. The first blast of wind was like a bomb exploding. It hit us so hard that it blew one of our sleds down the glacier slope. Helpless to do anything but pile on top of one another to keep warm and to prevent anything else from being blown away, we lucked out when there was a momentary lull in the storm five minutes after the first blast hit. That gave us just enough time to regroup, retrieve the sled and get back down to a more sheltered spot before the full force of the storm took hold.

Four days of confinement in a small tent with three men on reduced rations was almost enough to convince me never to do anything like that again. And yet the allure of the Cirque rising from the misty heights of this largely untouched wilderness was irresistible. So I asked Demuth if he would give me a little time to think it over.

Hard as I tried to learn more about these icefields over the next few days, I could not come up with anything that would help me make a decision. From what I could tell, these icefield clusters on the Yukon–Northwest Territories are part of a lost world. Even on the most detailed topographical maps, they are without a name. Nor are they inventoried in the *North American Atlas—Glaciers*. When I asked Parks Canada officials what they knew, no one could tell me much about them even though they were included in the park expansion plans. Only Chief Park Warden Steve Catto, a veteran who knew more about the country than most anyone else, had been anywhere near them.

Not that I was entirely ignorant of what I was getting myself into. I had spent two weeks at the foot of this icefield the summer before helping Demuth's team measure the retreat of ice downstream of the Butterfly Glacier, a tongue of ice that melts into Bologna Creek. Demuth and his wife, Margie, were higher up on the other side, planting long aluminum poles into the tongue of the Bologna Glacier[1] so that he could monitor the retreat in the future. Most of what I did was on solid ground. But the short time I spent on the ice was not enough to make me think twice about

1. Bologna and Butterfly are unofficial names that we gave these glaciers when we were camped at the site.

what I was getting myself into. As much as my mind and body warned me against doing this with three men who were six, twelve and twenty-five years younger than me, my heart begged to go. I couldn't resist.

Fort Simpson is a 1,056-mile (1,700 km) drive from Edmonton, Alberta, and not an easy place to get to in late April. The last few hundred miles take you along a narrow gravel road that is at best treacherous and occasionally closed to traffic when rains wash parts of it out. There was only one service station on that road, but it closed up for good the previous summer. Crossing the Liard River to get to Fort Simpson can also be a problem. Throughout the winter and early spring months, an ice road allows cars and trucks to make the river crossing. But once the spring breakup starts, the road is closed. It can take as long as three weeks to put the ferry in service. So my only guarantee of getting there at this time of year was by plane to Yellowknife, the small capital of the Northwest Territories, and from there to Fort Simpson.

Getting to Tungsten, the site of a remote mine that produces a rare metal used for light bulb filaments, can be even more problematic. The mine is nestled in a mountain valley at the headwaters of the Flat River on the Yukon side of the border where it is shrouded in thick clouds or walloped by spring storms.

The day I settled into Fort Simpson, a series of these storms began to pound the mountains. I was stuck for three days before the pilot finally got the helicopter off the ground. When the weather cleared long enough for us try to get to Demuth's camp, a series of intense snow squalls nearly forced us to head back. Once we got to Tungsten, I still didn't appreciate what it was that I was in for. It wasn't until the helicopter left the mine site and labored to get above and around the shoulder of Mount Mulhulland that I began to have serious doubts. Looking down at the icy landscape below, the jagged peaks around us, and the fog and leaden clouds slowly boxing us in, I was beginning to understand how Wexler felt when he first saw the Cirque. This was as wild and rugged as anything I'd seen in a mountain setting. And by the looks of the precipitous drop that would take us down into the Fairy Meadow, our descent would have to be done with ropes and pitons, not skis and poles.

Demuth clearly didn't see it the same way. "Looks good," he said, pointing at the icy abyss below and giving me the thumbs-up. "This is going to be fun." The grin on his face couldn't have been wider.

The sun was beginning to burn a small hole through the clouds by the time the pilot dropped us off and headed back to Tungsten to pick up Steve, Matt and the rest of our gear. From this vantage point overlooking the Bologna Creek Valley, very little reminded me of the place we'd been to the summer before. A fresh blanket of snow took the edge off the spires around Mount Apler and the Vampires Peaks in the distance. There was also no sign or sound of the big meltwater stream that roared down from the tongue of the glacier.

Waiting for the helicopter to return, I took a short walk around to look for tracks of the caribou, grizzly bear and other animals we had seen signs of the year before. The only hint of life was the shrill whistle of a rock rabbit, a pika, that was hunkered down somewhere beneath the snow on the hillside. Ordinarily, I wouldn't have paid much attention. But this animal was going to be thc focal point of my next trip to the St. Elias icefields, so I was curious to see how it lived beneath the snow. Although the pika was safe down there in its nest of dead grass, dried sedges and freeze-dried berries, it was clearly not happy when I started to tunnel down.

"Over here," Demuth said, waving me back to where he was studying the map. "I didn't get a chance to tell you about the change of plan we made back in Tungsten. Here we are just south of Mount Mulhulland," he said, fingering the starting point we had planned on. "There's Mount Harrison Smith overlooking the Fairy Meadows. The icefield is everything in between. Losing three days as we did to the weather means it's going to be very tight to get to where we had hoped to go in the short time we have. The forecast is for more nasty weather to come. So the Fairy Meadows, I'm afraid, are going to have to wait for another time."

Although disappointed, I wasn't mortified. The new plan would have us crossing half the icefield in stages and returning to base camp each evening. That would mean longer days, but it would spare us the torture of lugging all the food, tents, sleeping bags and heavy drilling gear on our backs and behind us on sleds. More important, it would remove the challenge of trying to find a way down into the Cirque side of the icefield without imperiling our lives.

Not that any of that mattered much to Demuth. Being part of a second generation of glaciologists working in the Arctic, he had been mentored by a pioneering group that included Fritz Koerner, Gerald Holdsworth and Garry Clarke. These men were utterly fearless in pushing the limits of mental and physical endurance and patience. Koerner was Demuth's senior at the Geological Survey of Canada and part of a small group of men who laid the foundation for the understanding of how climate change affects Arctic ice. In 1968, he set off with Sir Wally Herbert and two other British explorers on a scientific expedition that was described at the time as one of the "most audacious" ever attempted. The 3,600-mile-long (5,800 km long) ski and dogteam trip, aided by periodic supply drops from Canadian Forces aircraft, took them from Point Barrow, Alaska, over the North Pole to Vesle Tavleoya, a small rocky island off the north coast of Svalbard in Norway.

When they completed the trip on May 30, 1969, British prime minister Harold Wilson praised their achievement as a "feat of endurance and courage, which ranks with any in polar history." Prince Philip, the group's patron, called it "one of the greatest triumphs of human skill and endurance."

Gerald Holdsworth's spirit of adventure was and still is equally audacious. In 1974, Holdsworth came up with the idea of using Mount Logan, the highest peak in Canada, as a promising ice-core site for paleoenvironmental studies. Given the danger and logistical nightmares associated with getting to and drilling at 16,400 feet (5,000 m) above sea level, few people gave him much chance of succeeding in getting the funding he needed. Not only would the drilling team have to spend weeks climbing to get that high, they would need pilots that were experienced and daring enough to fly in on oxygen and pick up their core samples. Six years later, Holdsworth led a team up the mountain and drilled the first ice core. When he returned the following year to finish the job, Demuth, just twenty years old at the time, was there to assist.

With his bleach-blond hair, blue eyes and baseball cap, Demuth looks very much like the climber that he is. There's a swagger in the way he walks and an attitude in the way he talks that begs the word *dude*, one of his favorite forms of salutation. But once the subjects of glaciers, snowpack and ice cores come up, the scientist within him takes over with a stream of metaphors for glacial processes—"mass balance falling into disequilibrium," "glacial budgets wasting away," "ice ablating into thin air"—each one of which could be wonderfully descriptive if not always comprehensible to a layman like me.

"The best way to describe it to someone who wants to know why we bother worrying about melting glaciers is to liken them to water towers," Demuth said as we searched for a hard flat spot on the snow to set up our tents. "About seventy percent of the world's freshwater is stored in glaciers. Some of this is released during periods of drought and intense heat when water supplies are in short supply. If these waste away, we lose whatever insurance we have stored in that ice. There are other reasons why they're important. They can influence weather and climate. Certain forms of wildlife depend on them. They contribute to rising sea levels."

The fact that this icefield in the Northwest Territories is retreating is hardly earth-shattering news. Most of the world's glaciers are in retreat. Climate change is melting the European Alps, the snows of Kilimanjaro and the glaciers of Patagonia in South America. Thirty years ago, there were 156 glaciers in Montana's Glacier National Park. Today, there are just 26. Of the 850 glaciers that Demuth has been monitoring on the eastern slopes of the Rockies, 325 have disappeared entirely since the early 1970s.

The meltdown occurring here in the Northwest Territories, the Yukon, Alaska and in other parts of the Arctic has been more spectacular than anything scientists are seeing elsewhere because the climate in the polar world is warming faster than it is anywhere else. In Alaska, where glaciers cover 5 percent of the state, 99 percent of them are retreating. The Columbia Glacier at Prince William Sound has melted back 9 miles (15 km) in the last twenty-five years. Of the nineteen glaciers in the state's Juneau Icefield, eighteen are receding. The situation is similar in the southwest corner of the Yukon, where one of the world's largest subpolar icefields manufactures several glaciers that are 44 miles (70 km) long and more than half a mile (1 km) wide.

In the mid-1960s and 1970s, the Steele was one of a number of surging glaciers that tore through the St. Elias Mountains with a force that dazzled even the most seasoned glaciologists. At one point, more than 1.5 billion tonnes[2] of ice was being churned up and pushed forward at a rate of up to 50 feet (15 m) per day. But these days, this so-called "Galloping Glacier" is too wasted to make another run of it. Like its bigger sister, the Lowell Glacier, which surged up against Goatherd Mountain 155 years ago, it is in retreat. Now there is very little likelihood that either one will surge across and dam the Alsek River, as the Lowell has done several times over the past thousand years. Had the Yukon town of Haines Junction been around in 1850, the last time the Lowell butted up against Goatherd Mountain on the other side of the Alsek, parts of it would have been submerged under water.

As quickly as many of the continent's glaciers are melting, there's still more than 96,525 square miles (250,000 sq. km) of ice locked up in the glaciers and ice sheets. Three-quarters of that is located on the Arctic Islands; most of the rest can be found in the Canadian Rockies, the northern interior of British Columbia, the southwest corner of the Yukon and the coast of Alaska. Some of these ice sheets are enormous and will remain so for tens of thousands of years, even as the climate continues to warm. The Juneau Icefield, for example, extends through an area that covers 1,506 square miles (3,900 sq. km), the Stikine Icefield covers 6,500 square miles (16,835 sq. km) and the Wrangell–St. Elias Icefield is a sheet of white that spreads over 3,089 square miles (8,000 sq. km).

---

2. Metric tonnes will be used throughout; 1 metric tonne equals 1,000 kilograms, or 2,204.6 pounds.

The melting of Arctic and sub-Arctic glaciers has huge implications for low-lying coastal communities everywhere. Greenland has 694,984 square miles (1.8 million sq. km) of ice that is on average 1.4 miles (2.3 km) thick. If it were to melt completely, ocean levels would rise by up to 23 feet (7 m). That's not going to happen any time soon. But scientists at the University of Colorado, Boulder, point out that glaciers and ice caps, more than Greenland or Antarctica, are responsible for the ice in the world's oceans. They estimate that the melting of ice in both places is contributing about 100 cubic miles (417 cu. km) of ice each, which is as much water as there is in Lake Erie.

Not only has this flow of ice been steady, the output has been rising by about 3 cubic miles (12.5 cu. km) per year. If it continues to rise at that rate, many of the 104 million people who live within 3 feet (1 m) of sea level would have to move by the end of the century. Beyond that time frame, the entire island of Manhattan would eventually be threatened. The possibility is no longer just theoretical. That small rise in sea level has already forced the Alaska government to consider evacuating the entire Inupiat village of Shishmaref, located on a tiny island at the edge of the Arctic Circle. It's also why four other coastal communities—Kivalina, Koyukuk, Newtok and the Inuvialuit community of Tuktoyaktuk in Canada—may have to be evacuated as well within the next fifty years.

Compared to these behemoths in the southwest corner of the Yukon and Alaska, the icefield cluster along the Northwest Territories–Yukon border in the South Nahanni River Valley are tiny. The one we were about to tackle is no more than half the size of the city of Ottawa. It appears as little more than a thumbprint on topographical maps. Wrangell and St. Elias, by way of comparison, have their own page.

These small glaciers are more sensitive to climate change and therefore more reliable indicators of how the world is warming. Fritz Koerner, Demuth's mentor, likens these glaciers to canaries in the coal mine, an early warning, he likes to say, to the rest of the world of the consequences that climate change is bringing.

Koerner is also fond of comparing glaciers to museums. Hidden in the layers of ice, he often tells people, is a record of every atmospheric event of the past—summer melts, cooling trends, volcanic activity, industrial emissions, even bomb testings. The deeper you go down, the farther you go back in time. The more sophisticated the technology you use to interpret the cores, the more detailed the information you retrieve.

Since 1980, when Holdworth's group extracted an ice core on Mount Logan that was 336 feet (102.5 m) long, scientists have been drilling deeper and deeper into glaciers in the hopes of extrapolating an ice core record that will place the present climate, warmed by the burning of fossil fuels, into an historical context. Holdsworth's first core yielded data going back four hundred years. The 568-foot (173 m) core that he, Demuth, Koerner and others pulled out a decade later is still being analyzed, but what's being learned may eventually tell a story that is ten thousand years old.

As the icefields, glaciers and ancient snow patches of the world succumb to the heat that is being pumped into the polar and subpolar environs, scientists are reaping insights into the ancient past in unexpected ways. When caribou biologist Gerry Kuzyk and his wife, Kirstin, were hiking along an icy hillside in the southern Yukon in the summer of 1997, Kirstin's curiosity and keen sense of smell led them to a large mound of caribou pellets spread out around some caribou antlers sticking out of the snow. Normally, a find like this wouldn't draw much attention in the Yukon. In the warm months of summer, caribou often take refuge on the snow and ice to escape the swarms of flies that make life miserable for them. But the last time anyone reported a caribou anywhere near Kusawa Lake in this part of the Yukon was in 1932. Even more extraordinary was the size of the deposit. It was more than 980 feet (300 m) long, 644 feet (200 mm) wide and ankle- to knee-deep in some places.

The Twilight Zone is how Yukon caribou biologist Rick Farnell described it when he, Canadian Wildlife Service scientist Don Russell, Yukon government archeologist Greg Hare and others eventually arrived on the scene to investigate. Buried in that spectacular pile of dung, they discovered over the next two years, were the remains of freeze-dried birds, a rotting muskrat and the arrows, darts and spears that would have been used to kill them. Radiocarbon dating by several labs in North America estimated the dung and artifacts to be up to nine thousand years old. What the Kuzyks found melting out of that ancient swath of snow and ice that day was the accumulation of everything that had either died or been deposited in that time.

Greg Hare was certain he would never see anything like it again. But two years later when he was still working on those specimens, three sheep hunters came into the Beringia Research Centre in Whitehorse to inform officials of a carved walking stick and the remains of a man they found lying at the foot of a glacier along the B.C. side of the Yukon border. When leaders of the Champagne and Aishihik First Nations were informed of the discovery, they instructed the Yukon and B.C. governments to investigate the next day. Hare was among the first to be flown to the site. He was there as well when forensic anthropologist Owen Beattie was brought in to help remove the headless torso that was still partially embedded in the ice and transport it and a ground squirrel cloak, cedar woven hat and a pouch containing the remains of fish back to Whitehorse.

As it turned out, Kwaday Dän Sinchi (Long Ago Person Found) was not as old as archeologists initially suspected. Forensic examinations and a series of radiocarbon dating suggested that he was in his twenties when he walked the region one hundred and fifty to three hundred years earlier, well before the first Europeans ventured into this part of the world.

This wasn't the first time human remains and other artifacts have melted out of the ice and snow, and it won't be the last. In 1991, the body of a hunter who died fifty-three hundred years ago was discovered in the Alps. Ötzi, the so-called Iceman, became a worldwide subject of interest and a symbol of how climate change is manifesting itself at mid-latitudes. And since Kwaday Dän Sinchi opened the international scientific community's eyes to the possibilities of new lines of research, archeologists in the Yukon, Scandinavia, Italy and Austria have been actively monitoring snow and ice patch areas for new discoveries. This new field of study has become so fashionable that more than 130 scientists attended the first snow patch conference in Switzerland in 2008.

Of the eighty to one hundred snow and ice patches that Hare and his colleagues are keeping an eye on in the Yukon, twenty-three have produced important archeological artifacts. Some of the weapons they've found are so perfectly preserved that they still have organic material and blood on them. This has allowed Hare and his colleagues to resolve the debate about when people here started using bow and arrows instead of the more primitive darts. The dating of all these artifacts clearly show that dart throwing began around eight thousand years ago and ended around the time Mount Churchill in the Wrangell–St Elias went through a series of eruptions twelve hundred and fifty years ago. They theorize that ash and fallout from the volcano made the region so uninhabitable that people from the interior moved to the coastal regions or elsewhere. There, they acquired the bow-and-arrow technology before eventually moving back.

It's often been stated that the true art of climbing a mountain of snow and ice is confronting the inherent dangers with a healthy dose of prudence. Demuth has no toenails to remind him of that. He lost them in 1981 while ascending Mount Logan on the expedition with Gerald Holdsworth: on a particularly cold night, he forgot to put the inner liners of his climbing boots into his sleeping bag. He never managed to thaw the liners out after that.

Alpine glaciers present other formidable challenges to climbers, especially when they have heavy drills, aluminum poles and shovels strapped to their backs. Many glaciers are in a violent state of flux, retreating or advancing in response to changes in precipitation and temperatures. When a glacier slides over ice falls or makes a turn around a valley bend, gaping fissures often form under the strain. These so-called crevasses can be tens of yards deep and difficult to detect because they are covered in snow for most of the year. Because the opening in a crevasse is generally wider at the top than it is at the bottom, the chances of getting wedged in partway down before hitting bottom is a distinct possibility if one falls in. If internal bleeding doesn't do the victim in before rescue, hypothermia likely will.

No doubt Steve Bertollo had this in mind when it was left to him and me to find the poles the Demuths had buried into the glacier the summer before. Perhaps it was my relative inexperience climbing glaciers or maybe it was because I was the senior citizen on this expedition. But his instructions that day were clear.

"If I fall through a crevasse," he said shortly after we roped up and switched on our avalanche beacons, "then you do your best to hang on. I'll get myself out. If you go down, I'll come and get you."

When I first met Bertollo in Tungsten, "gonzo glaciologist" was the impression that came to mind. With his long locks, wool toque and glacier glasses, he looked more like a hucker interested in finding fresh powder to ski than exploring the science he was hired on to do. It didn't take long, however, to figure out why he's been a go-to guy for veteran glaciologists such as Demuth and Gerald Holdsworth, with whom he has also worked. Bertollo has what every glaciologist needs for support—outdoor skills, a solid scientific grounding and a workhorse ethic. He also has a wicked sense of humor.

Although the sun was burning bright when Bertollo and I began our ascent late in the afternoon, the billowing clouds of vapor rising from the icefield suggested the trip would not be uneventful. A tempest rising from the South Nahanni River Valley was also promising to bring more misery. It was hard going, trudging up the steep slope of the icefall on short parabolic skis, which were new to me. After having spent so many years on long Telemarks, I had trouble figuring out the bindings' locking and swivel mechanism and getting into a comfortable rhythm. The going got a little easier once we got on top. But still, there was no opportunity to go slow and to acclimatize myself to the challenge at hand because I was roped to Bertollo and he was setting the pace, which was relentless despite the heavy ice-coring drill on his back.

This brisk pace was a minor annoyance compared to the swirling waves of snow that eventually hit us. Within the hour we were in and out of the clouds and snow, wiping the sweat pouring off our foreheads one minute and then pulling up our hoods the next trying to stay warm in the soupy moisture.

It was also getting to be late in the day, although in the midst of a whiteout, one abandons the prospect of night and day because there is nothing to suggest a difference. The ghostly gray that shifts ever so slightly from dim to dark is all that there is to signal a horizon. These can be trying times for seasoned alpine skiers and uneasy ones for those who have not done anything like this before; when up from down means nothing and when safety depends on a GPS device pointing the way. All sense of progress was lost, with no sun to aim for on the glacial plateau. The human mind, reacting like a computer overloaded with demands, threatens to crash.

Marching along like blind men, we stopped only occasionally to catch our breath and dust off the thin layer of ice crystals glued to our clothes. It was snowing and blowing so hard our fresh tracks virtually disappeared with each short glide. Up there on that glacier, we were more ghosts than human.

Bertollo was apparently unfazed by all this. He finally stopped and casually planted his poles in the snow. "If this GPS is right," he said, taking a reading of where we were standing, "then Mike and Margie's pole should be somewhere right around here." And there it was, sticking 3 feet (1 m) out of the surface, just a few feet away from where we stood.

I had once spent a summer digging ditches in hard clay, so I was in my element shoveling deep so that Bertollo could get a sense of how much precipitation had fallen since the Demuths drilled and planted the pole eight months earlier in the so-called zone of ablation on the glacier. Accumulation and ablation are the two opposing processes that glaciers are always undergoing. Accumulation is the process in which new ice is formed. Ablation is the process in which ice melts. When the amount of ice formed in a season is the same as that which melts, the glacier is in a state of equilibrium. Now that climate change has been warming up the polar and sub-Arctic regions, most glaciers have been falling into disequilibrium; more often than not, the snow that builds up on top of a glacier cannot keep up with the melting and calving that occurs farther down.

Aluminum poles drilled into the ice in lines that go up and down and across the glacier serve as reference markers that help glaciologists measure seasonal changes and the movement of ice. Chest-deep in the hole we dug, Bertollo took several samples of snow at various levels along this pole line to weigh and measure their density. By doing this, Demuth can later determine how much snow fell and how much melted in those months since he and Margie first planted the pole.

In some cases, glaciologists can measure the retreat of a glacier by studying vintage photographs. This was one of the ways Demuth and others measured the retreat of the Peyto Glacier in Banff National Park several years ago. Using a *National Geographic* photo taken by explorer and photographer Walter Wilcox in 1898, Demuth was able to deduce that the ice had retreated more than 6,500 feet (2,000 m) since then.

Topographical maps are also useful in delineating retreats because many of them were published decades before the current meltdown got under way. The map of these icefields, for example, was produced in 1950, five years before Wexler and his fellow climbers first ventured into the Cirque. It clearly shows that our base camp would have been buried beneath a thick tongue of ice that extended several miles down into the valley. These days, satellite imagery helps glaciologists delineate glacier cover far more accurately than any photo or topographical map.

Yet, as many tools and techniques as there are available to glaciologists, forecasting the timing, duration, magnitude and impact of glacial surges, retreats and runoffs is still a complicated exercise. So many factors influence the behavior of a glacier that it's almost like some of them have minds of their own and don't bother following the basic rules of physics.

Glaciologist Garry Clarke and his University of British Columbia team found this out in painstaking fashion during nearly forty years of research in the St. Elias Icefields. In 1980, Clarke and his group detected major changes in the flow of the Trapridge Glacier that suggested it was about to surge forward. This tantalized them for more than two decades because the glacier kept on advancing.

For as long as they stood by and monitored the situation with sophisticated sensing equipment, the Trapridge never switched into high gear. Clarke was not completely disappointed because a huge amount of crucial scientific data was collected as they waited for all hell to break loose on the ice. But in the summer of 2007, Clarke closed down the Trapridge project, still puzzling over the implications of what he saw. He suspects that accelerated climate warming of the past decades has thrust the glacier into a losing situation. The glacier, said Clarke, was trying its best to surge but lost the battle against simultaneous mass loss by summer melting.

Over in Greenland, the situation can be just as complicated. Coastal glaciers there are melting into the Atlantic Ocean twice as fast as previously believed. But snow and ice have also been building up in the interior. This has led climate change skeptics to wrongly make the claim that the ice sheet is not thawing out. The reality is that warming can both trigger an advance or a retreat, depending on the thermo-mechanical nature of a glacier. Sometimes it's the amount of water underneath the glaciers that helps propel it forward or backward. Other times it's the weight and pressure of snow that builds up or sublimates at higher elevations. More often than not, it's a combination of both. Thanks to radio echo data and ten years of radar information, scientists have recently confirmed that the Greenland Ice Sheet is, in fact, slimming down dramatically. The data shows that the annual loss of mass has risen from 22 cubic miles (90 cu. km) in 1996 to 36 cubic miles (150 cu. km) in 2005.

We were not long on the glacier when the pale yellow sun began to set behind us. The warm air that had been so enervating following the passing of the storm was now pricking us with a wintry sting. We began silently carving a new path through the fresh snow that had fallen on the glacier and along the narrow side valley that led back to base camp. On the way down, the skies were clear and the winds were calm. The only sign of life were two black dots that mirrored our every move on the other side of the valley. Stung by that same breath of cold air, Demuth and Beedle must have got the same idea and were going home as well.

Back at camp, I was more than eager to eat but a little uneasy about what would be on the menu. When suppertime arrived on nights like this the summer before, Dan McCarthy, the lichenologist who was part of Demuth's team, would invariably dip into his old camp-school recipe book and dig out a package of freeze-dried sweet and sour pork or a can of whole chicken, the gelatinous likes of which I thought disappeared from store shelves after the Great War ended. Hungry as both glaciologist Christian Zdanowicz and I were after spending eight to ten hours a day measuring the length and width of rock lichen on that trip, there were nights when we were both tempted to forego the feast that was enthusiastically offered. "You think this is bad?" McCarthy would say to us indignantly. "You should go out with my mentor, Brian Luckman, if you want to complain about bad food. Spork and Spam is just about all he eats when he goes out into the field."

Although I was surprised when I learned that Bertollo was a vegetarian, his tasty meal of fried falafels that night was better than I had hoped for. My only concern was that his breakfast menu was going to be caffeine-free, a worrisome prospect. I had not started a day without at least one big cup of very strong java in more than twenty-five years. Taking note of my disappointment, Bertollo warned me about what I might expect on my next trip to the St. Elias Mountains, where scientist David Hik was engaged in a long-term study of a colony of pikas living on the *nunataks*[3] in the icefields. "Kieran O'Donovan, the graduate student that runs that camp, is a lot more hard-core than I am about being a vegetarian," he told me. "Don't even think about bringing in meat or cheese. And there won't be any coffee either."

We were not long in eating. With the last rays of the evening sun lighting up the peaks with a rosy alpenglow, I pulled out my thermometer to see that the temperature had plunged to 5°Fahrenheit (-15°C) now that we were in the darkening shade. Chilled but not uncomfortable, I offered to do the dishes while the others prepared to slip into their sleeping bags in the two tents we brought along. With leftover herbal tea with which to wash them, snow was the only other cleaning agent available for me to do the job.

Following that painful chore, I thawed out my hands inside the arms of my parka and sat out in the open to contemplate our situation. Down below and all around us, there were places with ominous names: Broken

---

3. *Nunataks* are mountain peaks completely surrounded by glacial ice.

Skull, Headless Valley, Murder Creek, Deadmen Valley, the Funeral Range, Hell Creek, Vampire Peaks, Cathedral Creek and Pulpit Rock. Part of the naming, no doubt, had to do with the way these glaciers shaped this country. Viewed from a certain angle in the distance, the Vampire Peaks look remarkably like giant bats waiting to launch into the night. Pulpit Rock is unmistakable, resembling a giant lectern.

The Nahanni country is steeped in mythology. Since the headless bodies of Willie and Frank McLeod were found downstream along the banks of the Nahanni River in 1908, rumors of lost gold and untold wealth piled high in the valleys below these icefields were never ending. In the decades that passed, more than thirty-six people went missing. Some drowned, others died in cabin fires and one prospector who overwintered at Glacier Lake not far from where we were camped blew himself up with dynamite. When the RCMP investigated, they concluded that he took his own life out of despair that he was never going to make it through to spring. Legends of murder, of course, persist but remain unproven.

Many of these stories are well known to outdoor adventurers who come from all over the world for the chance to climb the Cirque or to paddle down the South Nahanni, the Little Nahanni or the Flat rivers. What's less appreciated is the rich history of the people who actually lived here year-round. For the Dene of the Nahanni, the geological features have different meaning. Downstream in one of the hot vents that rise up from an assemblage of hot springs along Rabbitkettle River, for example, the spirit Ndambadezha is reputed to live. Hunters passing by this area in moose-skin boats used to stop and pay homage to this supernatural being that chases a giant beaver downstream, where its slapping tail could no longer overturn their boats. These hunters were constantly on the lookout for the Naha, a fierce tribe of mountain people that would raid their camps, steal their women and children and take whatever else was worth carrying. On those few occasions when the surviving victims mustered the courage and the people necessary to retaliate, they would descend on the Naha at night only to find a campfire burning. The Naha eventually disappeared, but the Dene believe their spirit still lives on in the huge rock tower that splits Nahanni's Virginia Falls.

For paddlers, the Nahanni is what Everest is to mountaineers and what the Cirque of the Unclimbable is to rock climbers—wild, breathtakingly beautiful, challenging and mysterious. As many as a thousand people will spend a small fortune for the opportunity to come to this part of the world each summer. The meltdown of the icefields above the river, however, is becoming a concern for Parks Canada officials because the runoff very much influences environmental conditions and human activities in the park. The more water that barrels through the river's four deep canyons, the more dangerous a paddling trip becomes. Rain is one determining factor, but so is the rate of snowpack and glacial melt upstream. So much water has been pouring off the glacier in recent summers that the South Nahanni River has remained in flood mode well past the rain and snowmelt period.

This isn't just a concern for paddlers and adventurers. Water from the Nahanni eventually spills into the Liard and Mackenzie river systems. The less flow there is to those big rivers, the more difficult it becomes for barges to ferry food and supplies to Dene, Gwich'in and Inuvialuit communities. The less water there is flooding into the huge delta at the end of the Mackenzie River, the more vulnerable the fish and wildlife of the region become. And as much as some might not wish for the prospect, plans for hydro-electric development on the Mackenzie are dependent on upstream flows of water.

During the night, our exhalations and perspiration condensed into the nylon fabric of our tent, producing a layer of hoarfrost on the igloo-shaped ceiling. Big flakes of ice came raining down on Mike and me as we fumbled around for the liners of our boots. This is the most hated time of any trip, when one is forced to vacate a warm downy nest for the wintry bite of a sunless morning. My fingers were numb by the time I got my liners and pants on. Looking at the thermometer, I wasn't surprised to find that it was -4°Fahrenheit (-20°C).

Outside, a thin line of light crawling slowly down from the highest peak in the distance suggested that it was going to be a while before the warmth of the sun would reach us. In the rush to prepare tea and breakfast, the only sounds made were the hard crunching of snow and the jet engine roar of the tiny single-burner stove heating up water. No eggs, bacon or toast here, just a few packages of instant oatmeal that we washed down with herbal tea. I tried not to think about coffee.

"No point hanging around here waiting for the sun to come to us," said Demuth. "We may as well get an early start and meet it halfway up the glacier." So we loaded up our backpacks with the drilling gear, an extra layer of clothing, lunch and bottles of water we would need along the way. It was going to be a very long day on the ice.

The sky had been cloudless all through the night and during the first hour of that morning, and it didn't take long for the sun's heat to begin to evaporate the fresh snow that had fallen the day before. Up ahead of us, streams of moisture billowed up like steam blowing from the depths of a geyser. It was tempting not to imagine, as the Dene and the Tlingit in the Yukon and Alaska did, that this glacier was alive—not only does it snake forward and retreat up and down the valley threatening to swallow us at any given point, it breathes as well.

Once again a tempest was moving in from the Nahanni Valley, portending more snow and whiteout conditions. This one came in much faster than the last, slapping us hard with a wet blanket of snow when we weren't expecting it. Ski goggles misting up with each exhalation of air, I could barely see Matt Beedle breaking trail just 6 feet (2 m) ahead of me.

Although it had seemed godforsaken at the time, it was glorious when the skies partially cleared an hour later. It was surreal up there on top of this plateau of ice and snow. Above us, the clouds were crashing into the *nunataks* all around us. Fresh snow, which had fallen pure white, was shifting color from cobalt blue to various shades of gray in the reflection of the sun and the shadows of the clouds.

With a caffeine-withdrawal headache coming on strong, I took a long drink, hoping that the throbbing pain was more the result of dehydration than being denied a cup of coffee in the morning. When I saw that I had emptied half the bottle of water, I realized that my thirst would have to wait to be quenched later. We had several more hours to go and a couple more holes to dig before the day was over. As much water as there was up here, we might as well have been walking across a desert. What there was available to quench our thirst required a stove to melt it.

In the short time we were up on that icefield that day, the wild weather never did let up. For every hour of scorching-hot heat reflecting off the white surface, there was another hour of wind, snow, sleet and fog. Whether this was just the glacier making its own weather or another signal of the variability that climate change brings was impossible to say. But I counted twelve weather shifts that day. All we had to guide us down at one point were the GPS and what was left of the tracks our skis made on the way up. So much snow had covered them that Beedle, leading the way at the time, suggested that maybe we really were blind men because it was, as he imagined, akin to reading Braille.

After nearly ten hours of this, it was no longer fun. Out of water, I was parched. Forced off caffeine, I had a throbbing headache. Plowing downhill for so long, my legs felt like rubber. I figured we'd all collapse into our sleeping bags by the time we got back to base camp. But just before we hit the icefall at the bottom edge of the glacier, I glanced over to see how Demuth, roped up behind me, was doing. He was zigzagging back and forth like a slalom racer going downhill in slow motion. It looked like he was just revving up for a good time. "Got to get in a few turns, man," he said, smiling broadly as he unroped and swished past me. "This isn't just science you know. It's an adventure."

And then just like that, he disappeared over the side of the steepest part of icefall, leaving me to contemplate how I was going to follow his tracks.

chapter three

CHANGING LANDSCAPES

– Kluane National Park, Yukon –

The weather never change that much years ago . . . it is always cold. Not like today. You can't even tell when the weather is going to change. Years ago, we know when the weather is going to change—mild weather it is going to get storm come, we get ready for it even. But today it changes so much . . . boy we expecting a big storm. Next day, clear as can be. I can't predict the weather anymore like we used to years ago.

—Peter Esau, Sachs Harbour, 1999

ON THE MORNING OF MAY 25, 2005, three men were coming to the end of a long, hard climb on Mount Logan, the highest peak in Canada. Although some of their fellow climbers decided to push on to the summit, Erik Bjarnson, Don Jardine and Alex Snigurowicz opted to follow the others who had already headed back down to take advantage of the clear skies.

Initially, the descent was routine and uneventful. But after the climbers turned a corner on a narrow ridge just above a plateau at 17,000 feet

(5,200 m), they were hit by a blast of wind that sprang out of the Pacific Ocean. Erik Bjarnson, a veteran climber, described the whiteout conditions as akin to entering the gates of purgatory. Realizing that they were perhaps in the worst place they could possibly be, they huddled together behind a rock and their tent fly to figure out their options. Don Jardine thought their best hope was to put on crampons, leave the skis behind and kick-step down to a more sheltered spot. But Bjarnson's fingers were numb by this point, so they decided to set up their tent and sit it out.

With the benefit of hindsight, it was the wrong thing to do. The storm worsened. Every three seconds or so, a blast of wind hit them, threatening to rip the tent apart. Later that night, the intervals between the blasts lengthened to six seconds, which Jardine thought might be signaling a waning in the storm. That wasn't the case; he eventually realized that the longer it took for each gust to come, the more powerful it would be.

Bjarnson's hands were seriously frostbitten by the time morning arrived. Realizing how vulnerable they were, Jardine and Snigurowicz decided it best to go out and build a snow shelter just in case the tent didn't hold. Moments after they stepped out into the storm and lightened the load that was holding everything down, an even more powerful blast pulled the tent off its moorings, flipping it over and dumping Bjarnson and his sleeping bag on the snow. Another powerful gust blew the tent and most of their supplies away.

Without shovels, Jardine and Snigurowicz spent the next several hours digging out a snow shelter with pots and pans. Bjarnson stood by helplessly without the gloves that had been lost with the tent. With two frozen hands that banged like rocks when he slapped them together, there was nothing he could do to help. Inside the shelter, the men wrapped themselves up in the two remaining sleeping bags and hoped for the best. Thinking that they might not make it out alive, Jardine wrote a good-bye note to his family.

After nearly twenty-four hours on that 33-foot-wide (10 m wide) ridge, their prospects for getting out alive were fading. Dehydrated and with low blood sugar, Snigurowicz was slipping in and out of consciousness and hallucinating. Bjarnson's hands were like bowling balls and

Jardine's toes were frostbitten. But several things happened that saved the climbers' lives. First, the skies cleared shortly before midnight on the eve of May 27, which enabled climbers they had been in radio contact with to come to them and set up a tent. Then the weather held out long enough for a Hercules aircraft and helicopters from the Yukon and from Alaska to fly in for the rescue. After they were finally released from a hospital in Anchorage, Bjarnson was lucky to have lost only half his fingers And Jardine was thankful that he lost only part of a few toes.

Storms are a fact of life on Mount Logan. Because the massif is close to the moisture in the Gulf of Alaska, cyclones can rise up from that part of the world at almost any time of year. But of all the storms that have battered these mountains and icefields in the months of May and June since 1948, only 1 percent equaled the energy that walloped those three men in 2005.

Scientist Gerald Holdsworth would come to learn that there was something different about this storm—and not just because he was on the mountain at the time it struck. He had just finished setting up an automatic weather station to go along with another that was already set up 3,900 feet (1,200 m) below. With the help of satellite images and data from the U.S. National Centers for Environmental Prediction, he and physicist Kent Moore were able to put together a synoptic profile of the storm in the months that followed. This, they discovered, had been a storm in waiting. It was already laden with moisture from the tropics by the time it got to the north Pacific. The moisture was also unusually warm and ready to explode once it hit the cold air. Holdsworth described it as a *marine bomb*, the term meteorologist Fred Sanders coined to describe explosively intensifying oceanic storms.

The increasing frequency of storms such as this and the more common ones that draw their moisture from the Gulf of Alaska may be signaling a shift in the kind of climate variability that many scientists say will come with global warming. With warmer waters and less water being locked up in the ice, the meltdown could be adding fuel to those storms that have more local origins.

This has not made life easy for climbers or scientists such as Holdsworth, whose goal it is to climb Mount Logan or to get to the vast icefields that surround it. Where at one time a two- or three-day weather delay was as much as might be expected, it's not unusual now for them to be sitting tight for a week, usually more. And so I found myself in this holding pattern when I traveled to the Yukon two years later. Three days into a plan that would put me onto the Hubbard Glacier, I was sitting tight at the Kluane Research Station waiting for the icefield storms to pass so that pilot Andy Williams could fly me in.

Not that I hadn't been warned. Scientist David Hik, the biologist whose camp I was heading for on the Hubbard, had told me to factor at least a week or more into my plans in case there were delays. I was already committed to an icebreaker journey through the Northwest Passage in early July, but I ignored the advice. I really didn't want to miss an opportunity to participate in a study of an animal whose time on these icefields may be coming to an end.

The collared pika is truly a remarkable creature. It looks like a mouse and very much like Pikachu, the popular Japanese Pokémon character, but it is no ordinary animal. Living at high elevations on *nunataks*, those oases of rock that barely peek out of the vast sheets of thick ice that cover much of the southwestern Yukon and Alaska, a small number of their colonies are surviving on what little grows there. Despite how well adapted they must be to have endured for so long on so little, they and their cousins in the Ruby Range near Kluane Lake appear to be in serious trouble. In the past decade as many as 80 percent of some colonies have collapsed, largely as a result of the same climate variability that is responsible for the storm activity.

Not all pikas live on *nunataks* and glaciers. Most, in fact, are found along the talus slopes of the mountainous regions of western North America, where they spend a good deal of their time under rocks and snow. The one curious exception is a 500-mile (800 km) stretch of apparently suitable habitat that separates the Rocky Mountain variety from the collared pikas of Yukon and Alaska. No one knows why they are absent from this area, or if that's the really the case. Hik suspects that no one has bothered to look hard enough.

Although the pika looks like a mouse, it is more closely related to the snowshoe hare, an animal that shares the practice of eating its food twice, once in plant form and another as feces. This strategy accounts for the compressed peppercorn nature of its scat. But unlike arctic hares, which will sometimes bunch up in numbers that make them a hazard to bush pilots landing on remote airstrips, pikas don't tolerate neighbors getting any closer than 65 feet (20 m) from their territory, which typically covers 4,300 square feet (400 sq. m). Not that there are dire consequences for offenders. Typically, the meeping sound of an outraged tenant is enough to frighten off an intruder. At worst, the pika will give chase.

Pikas can be hard on their offspring, sending them off to find new homes when the food supplies dwindle. This can be tough if you're a young animal on a *nunatak* with no idea where the next refuge might be on an icefield. Unlike red squirrels that chase their offspring away before winter sets in, pikas will at least wait until spring. That's probably why climbers and pilots such as Williams see tracks in the snow 40 miles to 60 miles (60–100 km) from the green belt and several miles from the nearest *nunatak*. Spring is the best time for science and adventure to be undertaken in this region.

Climbers have been reporting the presence of pikas on glaciers for more than a half-century, but no one really knew how these furry creatures could survive in a hostile environment that is snow-free for only about six weeks each year until Hik embarked on his project in the early 1990s to determine how climate change was affecting northern alpine ecosystems. Hik returned from those first forays into the icefields with insights that some fellow biologists initially found hard to believe. Not only did he confirm that there were healthy colonies living on these *nunataks* year-round, he discovered that some colonies were augmenting their diet by eating the brains of birds. Presumably, these were migrating birds that had died after being blown off-course by storms tracking over the region. Hik could tell they were feeding on the brains by the neat little holes the pikas chewed into the birds' heads.

This was such a popular item on some pika menus that Hik found bird carcasses stacked up like cordwood among some of their hay piles—a food cache that typically consists of plants and berries that grow on the *nunataks*. The dry, cold air and wind cures these plants, which the pikas "hay" and pile until they have enough food to get them through the winter. So long as the snow remains dry, these non-hibernating animals can manage quite well in these subnivean environments. When the snow cover becomes soggy, as it does when it rains or gets unusually warm, pika populations and their habitats collapse.

In a subpolar world that is being heated up by the warming Pacific and Bering seas, Hik has seen this happen twice over the past decade. It happened in 1999, when a warm winter and spring contributed to an 80 percent drop in the population in one study area. It happened again in 2000, when he and his students could find only eleven pikas. This year, I would later learn, was turning out to be just as bad.

This isn't the only way pikas are being affected by warmer temperatures. With a thick, furry coat that keeps their core temperature steady at 104°Fahrenheit (40°C), the animals literally blink out when that body temperature rises by a degree or more. Often, they die. The more the climate heats up in the Yukon and Alaska, as it's been doing in recent decades, the more vulnerable pikas are becoming.

It was late June and the winds that were blowing over the Kluane Research Station toward the icefields seemed to be bouncing off the Kaskawalsh Glacier back down the Slims River Valley. Dancing dust devils were whipping up silt and fine sand from the flats and sending them drifting in waves toward Sheep Mountain and Kluane Lake. Standing on the edge of the dirt airstrip near the shores of this lake, Andy Williams looked quite the picture when I went to see him. He was taking a long drag on a cigarette, watching a flock of swallows swoop across the sky. Before I could inquire about my prospects, a white-haired alpinist, a Dutchman who had been waiting for a week to be flown into those icefields, moved purposefully toward him, wondering out loud, and in not the most polite manner, why Williams wasn't fueling up his airplane. Like a horse that knows it's about to be saddled, Williams turned his head and looked like he might make a run for it. Instead, he did what he always does when confronted with adversity; he cheerily accepted the man's complaints and nodded his head in sympathy. After the man's venting was over, Williams took another drag on his cigarette and pointed to the clouds ripping across the sky above the swallows.

"It's a lovely day down here, I'll admit, Ari," he said in typical Welsh fashion, clearly pronouncing every letter of every word he spoke and punctuating each with authority. "But it's a different story up there, where the wind is screaming across the icefields. We won't be flying today and I suspect we won't be flying tomorrow, the way things have been going. This is the way it's been all spring. I'm afraid you're out of luck for now, Harry."

Williams then sauntered over to his office, a run-down trailer with a gaping hole in the ceiling where a skylight was supposed to be. The Dutchman looked set to follow him, but reconsidered, turned around and stormed into the camp's rustic dining cabin.

"They trapped one this morning," said Williams when I knocked on the open door of his office before walking in.

"Caught what?"

"Kieran O'Donovan, David Hik's graduate student, caught a pika this morning up on the glacier. It's about the only good news they've been able to relay in the last week."

Andy Williams has seen more than his share of strange things during a four-decade career that has taken him to Antarctica, northern Quebec, northern British Columbia and the Yukon, and even he found it puzzling when he first saw these critters scurrying across the glaciers.

"When I first came across them, I remember thinking, 'Okay, here's a creature about 5 centimeters [2 in.] tall with a forward vision of about 1 meter [3 ft.]: what's it doing a hundred klicks from anywhere sensible?' Then we found abandoned burrows, reoccupied several years later. 'Where were they coming from?' Non-scientific theory: on the edge of the range, Mum says, 'Look, son, there ain't enough food for all of us, so off you go west, young man.' They then set off in their thousands, and a few arrive at this awful place. I've seen raptors hunting this area for years, and they're not up there just for fun. Come to think of it, Cabot, Columbus and the Southern Tuchone did much the same thing."

Williams has seen more than just pikas and raptors on the glaciers. Over the years, there have been wolves, wolverines, sheep and even a cougar, an animal not seen in the Yukon until recently.

"Most of them, granted, were walking a straight line and evidently not hanging around," said Williams. "But we've seen female grizzly bears and their cubs on and around the edges of the glacier where, presumably, they're trying to avoid male bears. I know that there are flies up there as well but not higher than 6,000 feet (1,830 m). I know that because when they hitch a ride with me in the plane, they fall asleep at that altitude. Unfortunately, they wake up just as quickly once I descend. Truth be told, there's probably a lot more going on up there. There just haven't been enough people around to witness it."

Williams first came to the Yukon in 1973, when he was recruited by the Arctic Institute of North America to fly scientists and climbers into the St. Elias Mountains and Icefields. He and his senior partner, Philip Upton, had two Helio Couriers at their disposal. The small four-seat airplane has an outstanding short takeoff and landing capability that makes it ideal for landing in tight spots at high elevations. When Upton died in 1984 after a long career that made him a legend among scientists and climbers, Williams and his wife, Carole, took over the show at Kluane.

I first met Williams in passing when I worked in Kluane National Park and briefly again in the spring of 1986 when he flew me and three others up to the Lowell Glacier for a ski trip that ended three days later than it should have because of a spring storm. He made quite the impression that second time when he flew in to pick us up. In the time it took to get to where he landed on the plateau of that glacier, he was already leaning back on the plane, soaking up the sun and dragging on a cigarette. The mischievous grin on his face suggested there was no other place he would have rather been except maybe in a lawn chair in that very spot.

Part of this good nature, no doubt, has to do with the job. Scientists and climbers are grateful to find someone willing to fly them to elevations that few pilots could or would dare to go. They are even more grateful when he flies in marginal conditions to pluck them off those mountains after weeks away from home. Williams also has unshakable confidence and a sense of adventure that is reassuring. When I asked him what he would have done had his plane broken down at 17,000 feet (5,200 m) when he was transporting ice cores off Mount Logan for Gerald Holdsworth, he simply chuckled. "Well, we organized it so that if we did break down or if the weather did move in, we would have enough oxygen available to allow us to climb down to levels where we could breathe more easily."

The fact that Williams and his wife, and now daughter and son-in-law, put on a good show at the research station doesn't hurt his image either. The station is primitive, with outhouses for toilets, two showers for as many as forty scientists, and just two or three telephones for computers to hook up to, but the food is superb, the atmosphere is collegial and the stories Williams tells are delightfully entertaining and informative.

There's no shortage of bush pilots in the Arctic who are willing to give you their two cents' on what's happening in the world, but scientists listen when Williams talks. Not only has he participated in dozens of research projects over the years, he's also co-authored a few scientific papers of his own. One paper he co-wrote with Keith Echelmeyer, a professor of geophysics at the University of Alaska, Fairbanks, describes his observations of change on the upper environs of the Hubbard Glacier. These proved to be reliable indicators of the surges that dammed the entrance to Russell Fiord downstream in Alaska in 1986 and again in 2002.

"These surges can be pretty dramatic once they get going," he said. "I first noticed that something was happening while flying over a cluster of smaller glaciers that feed into the Hubbard. First it was a fissure that formed in the snow and ice. The next time, it was freshly churned-up earth along the medial moraines. Pretty soon, places that used to be good for landing were no longer safe. I figured something was going to happen down below. It just took a few months for the shock waves to travel 125 kilometers down [80 miles] to the Alaska coast to prove me right."

A natural-born skeptic, Williams is not one to make sweeping pronouncements. Nor does he like to talk in absolutes. "That," he tells me emphatically, "is the problem with you writers, always looking for absolutes when none exist." But he concedes that the alpine world that he first flew over in 1973 is a much different place than it is today.

"You see it in the tree and shrub line that are moving up the mountain slopes. You see it in the icefields, though maybe not as much. There's just so much ice and snow piled up at those high elevations that it's going to take a lot of warmth to put a dent into it. But certainly I see it down lower in many of the glaciers. There are a few exceptions, but most are thinning out pretty dramatically."

Back at the dining cabin, the Dutch alpinist was sitting across the table from Gerald Holdsworth and some American climbers who had just come in to check on the weather that was holding them up as well. Like the rest of us, Holdsworth was horrified to learn that the Dutchman was suffering from the first stages of Parkinson's, a disease that obviously does not favor anyone who might find themselves alone in an emergency on a glacier with no one there to help them.

On his next trip out, Holdsworth was planning on flying to the icefields to see whether he could find any sign of a plane wreck melting out of the ice and snow. This plane crashed in 1951, he told us. On board was the wife and daughter of Walter Wood, a scientist and adventurer who was among the first wave of researchers and climbers to be based out of the Kluane Research Station.

At the time, Wood and his son Peter were on Mount Hubbard when his wife and daughter flew in with plans to photograph the peaks in the region. After overnighting on the plateau of the glacier, their pilot took off the next morning in foggy weather. Two of Wood's colleagues, who were on the plateau at the time, were apparently shocked that the pilot had taken the risk. They later recalled that upon takeoff, the sound of the plane that had disappeared in the clouds had lingered and didn't trail off—a sign, possibly that the pilot was circling and in serious trouble. No one heard from them after that. And in spite of a massive search and rescue effort, no sign of the plane was ever found. Although most people think the plane crashed in Yakatut Bay, Holdsworth has circumstantial evidence that makes a compelling case for the plane crashing into a nearby mountain.

Talking to Holdsworth that afternoon, it was difficult to imagine that he had done his first trip into these icefields in 1964 when he was a graduate student at Ohio State University. He looked far too young and wiry to be that old. It was fate more than anything else that first brought him here to the Yukon, he told me. The original plan to go to Antarctica was waylaid by logistical problems. So a supervisor suggested he come to the St. Elias instead. He's been coming back every year since after making a name for himself in scientific circles with his daring extraction of an ice core from Mount Logan.

Holdsworth first proposed the idea of extracting an ice core from Mount Logan in 1974. The six-year delay in getting up there, he says, resulted in large part from other commitments and the time it took to design and manufacture a drill specifically for this purpose. "It took three years to design and build it, and more time to get it to the drill site," he said. "There were a lot of things to consider: how, for example, it would be built so that it could fit into a small Helio Courier."

Although Holdsworth was part of the team that came back to drill another core in 2000–01, he's clearly most proud of the 338-foot-long (103 m long) core that was extracted in 1980–81. The data provided him and other scientists with insights into climate change that were significant enough to be published in *Nature*, arguably the most respected scientific journal in the world.

"Kent Moore at the University of Toronto and Keith Alverson from Switzerland co-authored that paper. It was important because up until that time, the only reliable data we had to measure atmospheric changes in western North America were ones that did not go back any earlier than the late 1940s.

"The oxygen isotopes in undisturbed snowfall layers stored annually in that core provided them with a climate record that dates back to 1736. Those layers at the bottom half of the core showed that for the first 150 years, the amount of annual snowfall was almost static year after year. But around the time the Little Ice Age ended in 1850 to 1870, they could see clear signs that snowfalls were increasing, presumably because the atmosphere had warmed. That increase in snowfall was even more dramatic in the later years. Between 1948 and 2000, the rate of snowfall increase was four times compared to the levels that fell 1851–2000."

The fact that the climate is dramatically changing the landscape in and around the Kluane area has not been lost on members of the Champagne and Aishihik First Nations, whose traditional territory overlaps in this region. For years they've been telling scientists and Kluane National Park officials that the alpine meadows that were once filled with low bush cranberry, sedges and other tundra plants are being taken over by thick stands of shrubs and trees. They've also complained that the animals they hunt are no longer found in places where they were once abundant.

These observations, in part, were what inspired Hik to undertake his long-term study on northern alpine ecosystems. As it evolved, the study of pikas on the glaciers turned out to be just a small component of a much more ambitious scientific enterprise. In addition to base camps on the Hubbard Glacier, Hik set up field stations on the Front Ranges, the Ruby Ranges and in the Burwash Uplands around Kluane National Park. Since then, he and his students have been doing all manner of things, measuring tree and shrub line advances, studying the population dynamics of ground squirrels, the population dynamics of hoary marmots and the horn growth of Dall sheep.

So far, their discoveries challenge the conventionally held view that the changes brought on by climate warming are slow and steady. What they found instead is that, like glaciers that have too much snow and ice accumulating on top of them, landscapes such as those along tree and shrub lines can surge ahead fairly dramatically if the conditions are right.

In coming to this conclusion, Hik and Ryan Danby, a Ph.D. candidate who is now teaching at Queen's University, looked at tree rings in much the same way that Gerald Holdsworth and Kent Moore got their data from ice cores. Like the layer of ice that reflects the amount of snow that falls in a given year, the core of a tree trunk has a ring that reveals the rate of growth in a single year. In warm weather, trees grow quickly and leave a thick ring of wood in the trunk. In cold years, this growth slows and the tree ring is thin. By examining tree rings from hundreds of samples they collected in the Kluane region, Danby and Hik were able to date the year that trees took root, the rate at which the tree grew and the year the tree died. The record they got dates back to 1700, just a few decades earlier than the information revealed by Holdsworth's ice cores.

They found that up until 1925, the tree line remained pretty much static in the Kluane region, neither moving up nor down the slopes in any appreciable way. But in the decades after the climate kicked into a warmer, wetter cycle, the tree line took off in earnest. According to Danby, it was as if someone turned on a switch that changed the drivers of the ecosystem. From around 1925 on, the tree line advanced by as much as 280 feet (85 m) up the warm south-facing slopes while the densities on the colder north-facing slopes increased by 65 percent. Even on the plateaus, they found that the tree line was moving northward at significant rates.

This was one of a number of eureka moments that Hik and his students were looking for. It explained, in part, why Dall sheep, which were once abundant, seemed to be in decline. It offers clues as to why the Chisana caribou herd, which Alaska and the Yukon share, was not successfully reproducing. The alpine environment these animals favor was either shrinking or shifting in ways that was making it difficult for them to adjust. The finding also raised some serious questions about the future. If the advancing tree line keeps forcing Dall sheep, pikas and other alpine animals farther and farther up the slopes and into more polar latitudes, what happens when they run out of elevation and mainland?

After giving up on my chances of getting into the pika camp in the icefields, I looked into the possibility of chartering a helicopter and flying into Goatherd Mountain, which hovered over the edge of the icefields. Fortunately for me, David Henry, the Parks Canada biologist stationed at Haines Junction, was willing to oblige when I asked him if he'd be interested in making the trip. It was, after all, a selfish request on my part, one that required some bureaucratic paperwork that he probably could have done without. But Henry was the only chance I had of getting into

Goatherd. Because it is such a fragile alpine ecosystem and critical to the survival of the biggest herd of mountain goats in Kluane, parts of the mountain are off-limits to campers. It is also extremely difficult to get to without a helicopter—a round trip on foot through the heart of Kluane's grizzly bear country to Goatherd would take at minimum ten days, if not more. My only hope was getting someone such as Henry, who had the authority to be there, to share the cost of flying in and hike with me along the route that would take us back.

Although I had never met Henry, our paths had crossed several times in other ways. He had occasionally written for a number of magazines I had worked for and we knew many of the same people in the scientific world. Had I been anyone else coming up with the suggestion at any other time, he probably would have declined. But after an eight-year stint with Parks Canada at Kluane, he and his wife were packing their bags and going back to Saskatchewan; one last trip to a place that some think is the most beautiful in the sub-Arctic was an opportunity he didn't want to pass up.

Henry was sixty-three years old, more distinguished in disposition than the cowboy or outdoor adventurer image that many Parks Canada wardens and biologists exude. The job in Kluane, he told me, was one that he had coveted for some time. Those first years, he said, met his high expectations. Not only did he get to explore the wonderful world of Kluane, he also got the chance to engage in some groundbreaking ecological studies in Vuntut, the national park that lies on the Alaskan border in the northern part of the Yukon.

But the realities of the job began to slowly grind him down. Like most national park scientists working with pathetically small research budgets and a corporate mentality that was once, and still is to some extent, uninterested in the bold advice of scientists, he found himself overwhelmed by the challenges that climate change added to the job. It was a factor, he told me, in his decision to make an early departure.

It was raining lightly when pilot Doug Makkonen picked us up on the front lawn of the warden station, just as it had been doing for most of this cool, unsettled Yukon spring. But looking down on the forest we began flying over, I realized that no amount of moisture was going to green up the trees below. From one end of the Alsek Valley to the other, virtually every mature spruce tree was a ghostly gray.

"Have a look, anywhere you want," said Makkonen as he slowly did a 360-degree spin of the helicopter. "What you see here is what we have in almost every valley of this park. It's all dead wood from mountaintop to riverside. It's just sitting there waiting for lightning to fire it up."

No one thought much of it in the early 1990s when the spruce beetle began to seriously bore into the trees of the southwest corner of the Yukon and Alaska. The six-legged, quarter-inch-long (half-centimeter long) bug has been feeding on small patches of trees in this part of the world for thousands of years. But the cold winter weather that used to keep these bugs in check is no longer lasting long enough to be the killer it used to be. Six or seven weeks of -31°Fahrenheit to -40°Fahrenheit (-35°C to -40°C) cold was once the norm in the south and central regions of the Yukon. Temperatures nowadays descend to those depths for only a week or two at a time.

The devastation that has already taken place is unprecedented. At least 40 million trees are dead or dying in the Yukon. Tens of millions more in Alaska are kindling. The voracious feeding cycle that used to play itself out after three or four years has now gone on for seventeen. Everything from the Kenai Peninsula in southwest Alaska to Kluane and the Shakwak Valley in the Yukon has been hit hard.

"When that first patch of gray showed up in this valley, they should have gone in and cut those trees down," said Makkonen, reflecting what many people in the Yukon and Alaska think. "It would have stopped the spread of beetles and produced forty full-time jobs. But what does Parks Canada do? They form a committee to discuss the options. They've been discussing the options ever since. Now it's too late to do anything about it. This is like a freight train. Nothing is going to stop it now."

Doug Makkonen was by nature an imposing character. For more than a quarter century, the helicopter pilot has been making news around the world by defying the laws of gravity, rescuing skiers and climbers at heights that were once considered improbable. He set an aviation record when he became the first pilot to do a sling rescue above 16,000 feet (4,900 m). He won a medal of bravery for rescuing four Austrian climbers in distress at the 16,500-foot (5,000 m) mark on Mount Logan. He probably would have earned another one had the Alaskans not waved him off when they flew in to rescue the three climbers who got stranded on the same peak in 2005.

In 2001 and 2002, Makkonen flew more than one hundred flights into Mount Logan, helping Gerald Holdsworth, Mike Demuth and their colleagues drill an ice core into Mount Logan. His list of passengers includes movie actors, biologists and even pregnant caribou from the Chisana herd. The caribou were immobilized before being strapped to the backseat of his helicopter and transported to a secure compound where they would be safe from wolves and grizzly bears.

In the back of the helicopter, Henry was listening silently to Makkonen's harangue. This wasn't the first time he had gotten an earful from locals. But realistically, there was nothing he nor anyone else in Parks Canada or the Yukon and Alaska governments could have done to stop this assault. Not only are winters no longer producing those killer cold snaps, most summers are so warm and dry that twice in the last thirteen years, the boring bugs have been able to complete a two-year life cycle in one. That's compounded the number of larvae gnawing away at the tissue that transports nutrients from leaves to roots. A healthy tree can fight back by producing copious amounts of pitch that flush bugs out from beneath its bark. But its ability to do so is seriously compromised when there is no pause in the attack.

This may be just the start of even more cataclysmic changes to come. The proliferation of the spruce beetle is not only destroying the mature forests of the Yukon and Alaska, it's setting the scene for a massive forest fire that could trigger even bigger ecological changes down the road. With temperatures rising, lightning strikes increasing, and all this dead wood just waiting to be ignited, there is the potential for an inferno that will be so hot and intense, it could burn down into the duff and fry the spruce seedlings and poplar saplings that are needed for regeneration.

If that happens over a very large area, says Rod Garbutt, the forest health specialist who has been monitoring the situation for the past fifteen years, it could be a silent spring in the southwest corner of the Yukon and Alaska for a long time to come. Only a massive planting program, he told me, would likely bring back those trees after such a hot fire.

Flying along the river valley that afternoon, we could see in the strand lines carved into the talus slopes of the hillsides evidence of climatic catastrophes that have hit the area in the past. Each line represents the spot where waves from Neoglacial Lake Alsek lapped up against its shores. This lake rose and fell several times over the past thousand years after the surging Lowell Glacier butted up against the mountain we were headed for that day. Each time an ice dam blocked the flow of water downstream, a long lake would quickly rise up along its headwaters.

"Back in the mid-1980s, there was some talk of using dynamite to bust a whole in the dam if it flooded the valley again," said Makkonen. "But my guess is that it would take a nuclear bomb to smash ice as thick and wide as that. The toe of that glacier is well over 1 kilometer [half-mile] wide."

The Lowell is one of about two hundred surge-type glaciers in the icefields of Alaska and the Yukon. The last time it made a serious advance toward Goathherd Mountain in 1997, there really was talk of detonating an explosive device to break it up. But Garry Clark, the University of British Columbia glaciologist who suggested it in jest, doubts the Lowell will surge that far ahead anytime soon. There just isn't enough ice left on the glacier to form a high ice dam.

That doesn't mean that other surge-type glaciers in Kluane won't go for a gallop. Irrespective of whether the climate is warming or cooling, glaciers can advance for many other reasons, some of which are still a mystery to scientists such as Clark. While the Lowell is in full retreat, for example, the Tweedsmuir Glacier was showing signs that it was about to surge forward that summer.

The panoramic view of the icefields was as sublime as it was picturesque. For a good ten minutes after we landed on top of the mountain, Henry and I just stood there silently as the massifs of Hubbard, Alverstone and Kennedy slowly peeked through the thinning clouds in the distance. "This alone," said Henry, summing up a sentiment I shared, "is worth the trip."

This wasn't the first time I had been to Goatherd Mountain. When I worked in the national park, I had spent several days here exploring the region with only a pup tent for shelter. It was a sobering introduction to the realities of camping at the foot of a massive icefield. On the first night of that trip, the aluminum tent poles of my tent bent like sticks of licorice when a wind storm came sailing down the Lowell. My partner and I were forced to spend the rest of the week sleeping in the open air. We were lucky that it didn't rain or snow.

In some ways, Goatherd was just as I remembered it. Up above and behind us, a blanket of thick gray cloud seemed to be affixed to the icefield on top of the mountain. Several meltwater streams cascaded down the slopes, filling up a number of tundra ponds along the plateaus before they overflowed and continued the descent. The colors cast in the different types of ice compressed in between the medial moraines of the glacier were also still the same, alternating from black to gray, from gray to blue and blue to white. It's as if an army of artists had spent a lifetime painting it. It didn't look natural back then and it still didn't look natural to me now.

The glacier that presented itself on this day, however, was nothing like the one I experienced the first time I saw it nearly thirty years ago. My best guess was that the toe of the glacier had receded by at least 1 mile (2 km). Not only was it dramatically thinner along the edges, it was no longer calving huge blocks of ice with any degree of frequency. The tiny ice-choked lake I remembered more than a quarter-century ago was now bigger, wider and pretty much ice-free.

And yet the sight of this 43-mile-long (70 km long) river of ice snaking down from the icefields was still spellbinding. Standing on the edge of Goatherd and looking down below, it didn't take much to imagine the devastation that occurred when Neoglacial Alsek Lake last drained in 1850. Geologists estimate it took just two days for all that water to drain out. The initial floodwall that came cascading down literally swept away the forests as well as the First Nations people camped along the river.

Both oral and archeological evidence suggests that the Tlingit used both the Alsek and glaciers to migrate from the coastal regions of British Columbia into the panhandle of Alaska and the southwestern corner of the Yukon. Not surprisingly, mention of this tragedy downstream of Goatherd is recounted in the traditional stories told to American anthropologist John Swanton in the 1900s. Many of them also reflect the view that glaciers were living beings that could see, hear, smell, open up and swallow people if they were in some way awakened or offended. "In one place, Alsek River runs under a glacier," an elder by the name of Deikinaak'w told Swanton. "People can pass beneath in their canoes, but, if anyone speaks while they are under it, the glacier comes down on them. They say that in those times this glacier was like an animal, and could hear what was said to it."

That evening, Henry and I gave ourselves just two hours to explore the area independently. The necessity of making progress was more pressing than our desire to stay the night and see the Lowell at sunrise. With just five days' provisions and one extra day of rations for an emergency, our return time on foot was going to be tight. We also weren't exactly sure whether we were taking the proper route. Hard as Henry searched through the Parks Canada records, he could find only onc trip report that offered any clues. Vague as that was, it came with a missing link on the map that we ourselves would have to connect. So with rubbery legs that buckled a bit when we helped each other put on our heavy packs, we began making our way around the mountain.

Initially, we were optimistic that we would find a way around the mountain. All we had to do was take directions from our GPS. But as early as it was in the season, we underestimated the amount of snow that was still on the slopes. Nor did we anticipate that there would be so many boot-killing rockslides to traverse. After four hours of stumbling over rocks and occasionally sinking waste-deep in snow, we managed to go just 2.5 miles (4 km) that night, less than half the minimum distance we wanted to cover.

The sun was already behind the mountains by the time we got feasting on a pot of hot water that Henry filled with soup powder, dried noodles, a carrot and a few chunks of dry sausage. "None of this gourmet cooking for me," he told me rather proudly. Much as I craved something more substantial, I harkened back to that can of gelatinous chicken that Dan McCarthy had tried to feed me on a previous trip. This, I reminded myself, is gourmet compared to what we ate at the Nahanni camp. There was also much beauty in the moonlit vista to distract me.

Up on Goatherd Mountain, the marsh marigolds, arctic poppies, mountain avens and other alpine species that we saw proliferating on the slopes and plateaus the next day suggested that all was still well in this part of the park. But signs of the trouble coming were clearly evident when we finally made our way into the valley on the other side of Goatherd Mountain. This supposedly wide-open, alpine country would lead us to Bates and Mush lakes and eventually back home. But after a half-day walking through this valley, it became clear that the tree and shrub line advances that Hik and Danby documented on the other side of the park were already well established here. The farther down the valley we went, the thicker and taller the shrubs became. Bushwhacking through this tall, brambly mess was no fun, knowing that a grizzly bear might be coming in the opposite direction. It was also eating into what little time we had left.

The advance of the tree and shrub line isn't the only way in which climate change is transforming the alpine landscapes of the northern world. Here in Kluane, the impacts of warming can be seen almost everywhere, in the production of spruce cones that red squirrels feed on, in the deciduous vegetation that is creeping up into the alpine regions, in the warming lakes that may be killing off Kokanee salmon and in the glaciers that are melting out of our icefields.

More impacts are likely to arise. Over at the Toolik Research Station in Alaska, scientists have discovered that warmer weather is tricking some tundra plants to flower twice in summer, which is a dangerous strategy because it makes the seeds of the plant that much more vulnerable to frost. The warm weather has also altered the timing of the green-up on the caribou calving grounds on the north slope of Alaska and the Yukon. In the short term, this is likely to be beneficial to caribou cows that need to feed themselves so that they can produce fresh milk for their young. But in the longer term, the vegetation might mature and dry up when the animals most need it.

Botanist Greg Henry (no relation to David) of the University of British Columbia speculates that there may be no stopping the advance of the tree and shrub line into the vast tundra regions of the Arctic. Using miniature greenhouses to simulate what the Arctic world will look like as the climate continues to heat up, he and his colleagues are now predicting that spruce trees will take root on the treeless landscape of the Arctic Islands in just thirty years.

By the time that happens, Henry warns, many of the moss, lichen and sedge-covered regions on the mainland of the Arctic will have shrunk dramatically. High Arctic plants, he points out, do not adapt well to a rapidly warming world. Nor do they compete well with southern species that are now moving north to take advantage of the warmer conditions. What's really remarkable, says Henry, is how quickly all this can happen. From extensive studies he and others have done, they've found that Arctic plant communities exhibit a detectable response to warming after just two years.

Change is taking place so fast in so many places of the Arctic world that some scientists are now wondering whether the entire ecosystem is moving into a new state not ever experienced by humans. Some of them are warning that this path to a new state could become a vicious loop. The more warming there is, the more melting there will be of the permafrost. The more melting of the permafrost, the more carbon there will be released from the remains of all those dead plants that have, for thousands of years, been safely stored in the deep freeze of the earth. At some point along this loop, the tundra will be heating up rather than cooling off the rest of the world.

In the end, David Henry and I never did make it out of the Bates Creek Valley on foot. After four days of hoofing it, we were still several days from where we wanted to go. So rather than push on, we gave ourselves one more day to explore this alpine world before calling in the helicopter.

Toward the end of that last day, I climbed as high as I could on Goatherd Mountain to get one last look at the icefields and the lush alpine meadows below me. This I knew would all be here the next year and several years after that. But would it be the same, I wondered, when I traveled here again in ten or fifteen years?

chapter four

IN NORTHERN MISTS

– Aboard the *Louis St. Laurent* –

How beautiful and warm and pleasant it will be in that warm sea around the north pole whare thare will be found all sorts of life and sorts of sumer fruits, whare the sun shines 6 months in a year and then have 6 months of night whare man can sleep or rest from thare long sumers toil. What a lovely vineyard to live in. How we envy Captain De Long.

—Whaling fleet captain B.F. Homan, when he learned of an American plan to sail across the North Pole in 1879

ON AUGUST 13, 2005, a massive chunk of ice the size of eleven thousand football fields broke free from the north coast of Ellesmere Island, 500 miles (800 km) south of the North Pole. No one was there to see it, but the collapse of the Ayles Ice Shelf was violent enough to register as a small earthquake at a Canadian military base 160 miles (260 km) away. The speed with which the ice broke away was breathtaking. When Laurie Weir, an analyst for the Canadian Ice Service, noticed the fracture on satellite images the next day, she contacted Luke Copland and Derek

Mueller, two scientists who had been studying this seascape for several years. After poring over the images, they could see that the island of ice had floated nearly a half-mile (1 km) within six hours after calving. Over the next five days, the shelf traveled another 43 miles (70 km). By the time winter's deep freeze put an end to its voyage, it had floated more than 50 miles (100 km).

The collapse of ice shelves is common in Antarctica, where hundreds of sheets of land-based ice are sliding on and eventually disintegrating into saltwater seas annually. But the collapse of the Ayles Ice Shelf, one of only six in the Arctic at the time, was the biggest in the northern hemisphere in more than forty years. And coming in a summer that experienced the biggest meltdown on record to that date, both Copland and Mueller were beginning to wonder how much longer the other ice shelves would be around.

Nearly two years after the collapse of the Ayles Ice Shelf, I was about to embark on a voyage through the Northwest Passage to get a firsthand look at the accelerating Arctic meltdown; 2007 was turning out to be another record year and the remains of the Ayles ice island that had stalled in its southward drift looked as if it might be on the move again. This time, it wasn't just Copland, Mueller and their scientific colleagues who were interested in what it was going to do. Frontier energy companies were also watching to see if the ice island might be headed for the Beaufort Sea, where Exxon had just invested $650 million for an offshore oil exploration lease. The catastrophic *Exxon Valdez* oil spill still weighed heavily on people's minds. The ship had diverted from normal shipping lanes that night in March 1989 when it ran aground on a reef. The multibillion-dollar environmental nightmare that followed was the worst man-made environmental disaster in history, and it happened in a part of the world where ice is not the hazard or cleanup challenge it could be in some places farther north.

When I arrived on the east coast with plans to board the *Louis St. Laurent* icebreaker at Dartmouth Harbour, the skies were clear and the air was warm and humid. Carrying a -40°Fahrenheit (-40°C) down parka that was too big and bulky to stuff into my backpack, I soaked it all in, knowing that this might be the last time I would experience the moist, verdant pleasures of summer for some time. By the time I got back from this three-week trip, and two other Arctic trips that were to follow, there was a good chance the cold bite of autumn would be in the air.

Getting a berth on the ship hadn't been easy. The competition for space occasionally made available to writers and photographers is fierce. I got my ticket from Marty Bergmann, the director of Canada's Polar Continental Shelf Project. Having been chief of the Arctic Science Division for the Department of Fisheries and Oceans, Bergmann knew how to navigate through the bureaucracy of the Coast Guard world and pulled the necessary strings. He was also a good friend of oceanographer Eddy Carmack, the architect of this expedition.

Coast Guard officials had warned me that security would be tight. On the long taxi drive from Halifax airport, I pulled out my passport, birth certificate, driver's license and the invitation from the Coast Guard to board the ship. Arriving as I was on the eve of the voyage, I wanted to make sure there was nothing that would stop me from going.

"Go ahead, dear," said the nice lady inside the harbor security hut, waving me on before I could offer her my papers. She was white-haired, twinkly eyed and completely uninterested in what I had to offer. "You must be the last of them. They've been coming in all day and all evening. It's over there."

I wasn't sure how to respond. I had expected much more than this.

"You can't miss it," she said and waved me on again with typical East Coast hospitality. "The *Louis* is a big red and white boat."

And so it was. Among the Canadian Coast Guard's fleet of vessels, the *Louis St. Laurent* is a five-star hotel. It is 390 feet (119 m) long and 79 feet (24 m) wide. In polite company, she is the "Queen of the Fleet," In not so polite company, she's "The Joan Rivers of the Fleet." The *Louis* has had so many facelifts and makeovers since she was built in 1969, the comparison to the surgically enhanced American gossip personality was apparently too hard to resist.

The *Louis* is designed for power rather than finesse. The ship has five big diesel engines that can generate 40,000 horsepower so long as they're fed up to 21,000 gallons (80,000 L) of fuel per day. Not all engines need to be running when the ship is in open water or cutting through softer, first-year ice. But they are necessary in the old, multi-year ice that is several feet thick and hard as cement. That's how the *Louis* made an historic two-ship voyage to the North Pole with the U.S. Coast Guard's *Polar Sea* in 1994. That set the stage for Ice Station Sheba in 1997 and other ship-based investigations into climate change in the Arctic since.

The *Louis*'s powerful engines make a lot of noise. This I knew because just below my tiny, windowless cabin, the engineers were revving up all five of them that first night to make sure they were firing on all cylinders. That may have been why I didn't hear Ed Hendryks, the biologist from the Canadian Museum of Nature, when he knocked on my door asking for directions. Being the last one on board, he was just as lost as I had been when I tried to find my room in the labyrinth of halls and stairwells that seemed to go off in every direction except to the place you wanted to go.

Hendryks and I hit it off right away. He was about the same age as I was but more wide-eyed about the journey that lay ahead. The two of us also had enough in common to generate some meaningful conversation. He had been aboard the Coast Guard's *Nahidik* in the Beaufort Sea the same year I hitched a ride to Tuktoyaktuk after kayaking down the Mackenzie River to Inuvik. He also worked with Dick Harington, the legendary paleontologist with the Canadian Museum of Nature, with whom I had spent a week excavating a 4.5-million-year-old beaver pond on Ellesmere Island.

Hendryks was here to inventory life in the mud on the ocean floor. He was one of ninety scientists and students on board this and another ship on the West Coast also heading north that summer. Together, they were going to do a 9,300-mile (15,000 km) transect of the Arctic Ocean. The plan for the *Louis* was to sail north through the Northwest Passage via Davis Strait and Baffin Bay. The smaller Vancouver-based *Sir Wilfrid Laurier*, in the meantime, was going to sail around the coast of Alaska and circumnavigate the Beaufort Sea. Over the course of six weeks, scientists and students on board both icebreakers were going to home in on everything from temperature, density and salinity of the seawater and microscopic archeates to giant bowhead whales. The goal was to see how Canada's three oceans are responding to and driving climate change and how climate change, in turn, is affecting marine life in this and other parts of the world.

Even though it was certain that we would run into ice, ours was not expected to be a difficult journey. Since the Arctic started melting in dramatic fashion in the 1990s, a number of ships and smaller vessels have had little trouble plying through these waters. Traditionally, the heavy older ice that gets flushed out of the permanent ice pack in Viscount Melville Sound and M'Clintock Channel forced even big icebreakers such as ours to turn south into Peel or Prince of Wales Strait rather than through the shorter, deeper passage in McClure Strait between Banks and Melville islands. That began to change when the *Louis* made the transect with almost no effort in 1998. For the first time in memory, there was no ice choking this passageway. It was also the first time in history that any ship had made the transect from east to west.

The apparent ease with which most of these vessels have made the transect through the Arctic in recent years stands in sharp contrast to the difficulties that beset those who had come before the meltdown began in earnest. From the time John Cabot first set off from Bristol in 1497 to the moment when John Franklin disappeared with 129 crew members on board the *Erebus* and *Terror* three hundred and fifty years later, more than one hundred and forty vessels tried and failed to find a passage through this icy world. Not until Norwegian explorer Roald Amundsen's *Gjoa* steamed through Lancaster Sound in 1903 and entered the Bering Sea three years later was the Northwest Passage successfully navigated.

Amundsen's voyage, however, did little to persuade other sea-faring explorers to follow. The prospects of getting through all that ice were still so daunting and dangerous that over the course of the next sixty-five years only thirty-six full transits of the Northwest Passage were made. Most of these were done for purposes of sovereignty, military strategy, tourism or commerce. Just one had anything to do with science. As a result, what we've learned about the Arctic in the past several hundred years comes from a mix of myth and reality, and truth and error that is only beginning to be unraveled and clarified. It's not a stretch to say that we know little more about the Arctic Ocean than we know about the surface of the moon.

For this we can blame the Greeks, who got everyone thinking for far too long that the North Pole was warm and ice-free. The word *arctic* derives from the Greek "Arktos," which means "bear." Arktos was the world that lay under a constellation of stars known as the Great Bear. The Greeks believed it was populated by Hyperboreans, immortals who lived beyond the Rhipean Mountains. There, Boreas, the God of North Wind, breathed violent gales that could shipwreck sailors as far south as the Mediterranean. In this idyllic world there were no seasons. The trees produced fruit year-round, "Lyres crashed and flutes cried." Everywhere, wrote the poet Pindar, were "maiden choruses whirling. Neither disease nor bitter old age is mixed in their sacred blood; far from labor and battle they live."

As charmingly delusional as this may sound today, the Greeks were far from ignorant about the physics of the ocean. In 310 BC, philosopher Theophrastus devised the theory that the Mediterranean was formed by the inflow of water from the Atlantic. He was also one of the first to consider the mechanics of fire and wind and the movement of storms. Theophrastus's understanding of the marine world may have given Pytheas, the astronomer, the courage to set out twenty years later in search of this mythical polar paradise. Very little is known about his remarkable adventure, but surviving accounts tell of a trip that lasted some six years. Following a well-established route mapped out by Phoenician sailors, Pytheas sailed through the Strait of Gibraltar and up the coast of the Atlantic before coming upon the Hyperborean temple of Apollo (likely Stonehenge in Great Britain). From there, he sailed north to Thule, where the sun did shine, as legend suggested, for nearly twenty-four hours. Yet instead of finding maiden choruses whirling and orchards producing fruit year-round, he was greeted by barbarian farmers who had little to offer

From inside a small cage on the west coast of Hudson Bay, the author looks up at an eight-feet-tall polar bear that is trying to find a way in.

Climate change has already hit hard the polar bear population of western and southern Hudson Bay. The U.S. Geological Survey predicts that two-thirds of the world's polar bears, including all of those in Alaska and most of Canada's western Arctic, will be gone by 2050.

Barren ground grizzlies of the Arctic may do well in a warmer world in which they do not have to hibernate for as long as they do now. Here, University of Alberta biologist Mark Edwards reunites a tranquilized cub with her mother.

Here Canadian Wildlife Service scientist Ian Stirling (red hat) draws a blood sample from a polar bear captured in the Beaufort Sea. With a collection of samples that spans three decades, Stirling and his colleagues have the tools needed to determine how climate change and other factors are affecting the health of these animals.

Glaciologist Mike Demuth of the Geological Survey of Canada begins his ascent of the Brintnell Glacier along the Yukon border in the Northwest Territories.

Steve Bertollo weighs and measures snow and ice samples from the Brintnell Glacier, part of a small cluster of icefields in the Northwest Territories that are rapidly melting.

A scientist skis across the top of the Lowell Glacier nearly 43 miles (70 km) from its toe on the Alsek River.

The Lowell Glacier is one of hundreds of surge-type glaciers in Alaska and the Yukon Territory that are rapidly receding. The white line denotes the position of the glacier's toe in 1980 when the author first visited the site.

Bird's-eye view from the crow's-nest of the Louis St. Laurent *icebreaker of a nearly ice-free Northwest Passage.*

A helicopter carrying a Canadian Ice Service analyst prepares to land on the Louis St. Laurent *in the Northwest Passage.*

The Ilulissat Glacier and icefjord have been on UNESCO's world heritage list since 2004. In summer, the glacier produces about 20 millions tons of ice daily.

Oceanographer Eddy Carmack tosses a message in a bottle into the Arctic Ocean in the hopes that the finder will report the location of its discovery. Simple experiments like this help oceanographers monitor changes in ocean currents.

The Croker Glacier was named after British explorer John Ross, who wrongly reported in 1818 that Lancaster Sound, the gateway to the Northwest Passage, was blocked by a chain of mountains.

in trade. Pytheas abruptly put an end to his voyage north when he was stopped by something he described as a "sea lung." This was likely a convergence of thick fog, undulating sea ice and gray sky, a not uncommon Arctic phenomenon that makes it look like one is sailing into an abyss.

The truths discovered by Pytheas and the Vikings and whalers who followed were not sufficiently compelling to dispel the myths and hoaxes that continued to shape the way people looked at the Arctic. Faith in the existence of an Arctic Eden seemed to be unshakable. This was most evident in the early maps of the area. Among the most notorious was the Zeno Chart, which was first published in 1558. Created by Venetian writer and mathematician Nicolò Zeno, it accompanied a narrative of the voyages of his ancestors, Antonio and Nicolò Zeno. The story tells how the two brothers sailed into the North Atlantic in 1380. Along the way, they were blown off-course and shipwrecked on an island called Frisland. There, they were attacked and nearly killed by hostile natives before Zichmni, a northern prince, came to their rescue. After forming an alliance with Zichmni, the Zenos came to learn of a place called Estotiland, where an intelligent civilization ruled over a kingdom rich with gold and precious metals. In the two decades they searched for this land, the Zenos purportedly reached Iceland, Greenland (Engroeneland) and a place they called Drogeo.

By the nineteenth century, historians were in general agreement that these voyages never took place. Prior to that time, however, the Zeno Chart was seen as an invaluable reference for anyone with the inclination to sail into the North Atlantic to search for new lands or for a shortcut to the Orient. Gerhardus Mercator perpetuated this delusion with his famous world map of 1569. Renowned for its precision and calculations of coordinates, the map is often cited as one of the greatest cartographic works of all time. But in this conical projection of the Earth, Mercator incorporates fictional features of the Zeno as well as medieval Scandinavian visions of the North Pole. So convinced was Mercator of the existence of this open polar sea that his map depicts a maelstrom at the North Pole swirling around a mountain in the middle of four islands inhabited by Pygmies.

The stakes were so high for anyone who found a polar passage that would take European sailors to the Orient that the idea of an open polar sea was too irresistible. In a letter to Henry VIII, Robert Thorne pleaded for financial support for the search for this polar sea, noting that it would give England an advantage over its Spanish and Portuguese rivals by shaving 6,200 miles (10,000 km) off the journey to the Orient. Like other entrepreneurs of the time, the former mayor of the port city of Bristol was tired of observing fleet after Spanish fleet unloading valuable Oriental and East Indian cargo while his countrymen watched from shore. "To discover the planet, mankind would have to be liberated from ancient hopes and fears, and open the gateways of experience," he wrote. That gateway, he believed, was to the west and the north.

Not everyone was convinced of an Arctic Eden at the North Pole. "What heat can the Sunne yeelde to that place above whose horizon he is never elevated more than 23 degrees and a half, a verie cold winterlie heat God wotte," Thomas Blundeville wrote in 1589. But generally, geographers and scientists were more interested in formulating theories that would support the existence of an ice-free Arctic rather than try to refute it.

With this in mind, many scientists spent an inordinate amount of time trying to prove how the polar sea could be open rather than why it was more likely to be frozen in ice. The earliest of these theories was predicated on two compelling but false scientific assumptions about the Arctic Ocean. The first, put forth in 1580 by geographer William Bourne, suggested that the long summer days in northern latitudes could create enough energy to melt the ice around the North Pole and allow ships to pass through. The second, posited by British explorer John Davis and others, was based on the belief that ice could only be formed in freshwater. The farther away one was from the freshwater flows of inland rivers, the theory suggested, the more ice-free the Arctic sea would become.

Speculative theories such as these continued on for another two hundred years. John Cleves Symmes, an American military officer, took the concept to an extreme when he put forth his Hollow Earth theory in 1818. He speculated that one could enter into this wide-open cavity at either end of the poles by way of this warm polar sea.

Right up until the nineteenth century, several explorers, including Constantine Phipps (1773), David Buchan (1818), William Penny (1850–51), Elisha Kent Kane (1853–55) and Isaac Hayes (1860–61), headed off on their adventures, convinced that a far less sinister open polar sea would get them to the Far East. It wasn't until American naval officer George Washington de Long sailed through the Bering Sea in 1879, certain that he would reach a small ocean of open polar water warmed by a Japanese current, that the truth began to unravel the fiction. Trapped in ice for fourteen months, de Long and his crew were eventually forced to abandon ship, dragging three small boats across the ice. After finally reaching open water, one boat was lost in rough seas. Altogether, nineteen men, including de Long, died of starvation or drowning,

Like Pytheas before him, de Long was not entirely delusional. He had been inspired to go off in this direction in part by Matthew Fontaine Maury's discovery in 1857–58 of a warm current that flowed north and August Petermann's theory that a similar current, the Japanese or Kuri-Si-Wo, which flowed though the Bering Sea and kept the water surrounding the North Pole free of ice for at least part of the year.

None of these discoveries and theories, however, was done in any systematic way until 1872, when the Royal Society of Great Britain sponsored the Challenger Expedition. Led by British naturalist John Murray and Scottish naturalist Charles Wyville Thompson, the expedition was provided use of the HMS *Challenger*, a British Navy corvette that was converted into an oceanographic ship, complete with laboratories and other scientific equipment on board. Over the course of four years, they sailed 7,920 miles (12,750 km), made 492 deep-sea soundings, 133 bottom dredges, 151 open-water trawls and 263 serial water temperature observations. Nearly five thousand new species of marine life were discovered. This trip set the stage for a new era in Arctic science.

The *Challenger*'s journey and findings inspired a new generation of Arctic explorers such as Karl Weyprecht, Fridtjof Nansen, Otto Sverdrup and Roald Amundsen. These explorers looked at the Arctic with the kind of perceptual framework that Charles Darwin had adopted while exploring the Galápagos Islands during the mid-nineteenth century. Just as

Darwin regarded strange plants and animals he encountered as natural adaptations to a unique environment, so did the new generation of explorer-scientist as they made their way through the Arctic. In fairly short order, distinct Arctic geological and oceanographic phenomena such as polynyas, pingos and polygons came to be seen more for the qualities they possessed rather than for the ones they lacked.

"We saw that to the eyes of the oldest civilization in history and down through the whole of antiquity, the North lay for the most part concealed in the twilight of legend and myth. . . ." Nansen wrote in *Northern Mists*, a groundbreaking book that analyzed and explained the many flaws of Arctic exploration up until that time. "Through all that is uncertain, and often apparently fortuitous and checkered, we can discern a line, leading towards the new age, that of the great discoveries, when we emerge from the dusk of the Middle Ages into fuller daylight. Of the new voyages we have, as a rule, accounts at first hand, less and less shrouded in medievalism and mist. From this time the real history of the polar exploration begins."

Emboldened by this new philosophy and inspired by the effortless manner in which the Inuit traveled from one place to another, Nansen became the leader of this new generation of scientist-explorer and one of the first to seriously consider the physical properties of the Arctic Ocean. He pursued this passion in remarkable style. First, he sailed into Greenland waters with a sealing ship. Then he crossed the Greenland ice cap on skis before coming up with the idea of drifting to the North Pole in a boat that was locked in ice. (A similar plan was independently crafted by the Canadian Joseph Bernier, but he could never get the funding or government support to do it.)

Nansen was intrigued by Norwegian scientist H. Mohn's theory of a current of ice and water exiting the Arctic Ocean between Siberia and Greenland. Nansen's own examination of an assemblage of microscopic plants, known as diatoms, that he brought home from his voyage to Greenland convinced him that there was something to Mohn's theory. Diatoms are found all over the world, but some species are exclusive to certain parts of the globe. Under a microscope, Nansen could see that this species he brought back from an ice floe off the coast of Greenland had its origin in the Bering Sea.

The discovery of objects belonging to de Long's expedition on the southwest coast of Greenland in 1884 merely confirmed what Nansen and Mohn speculated. There was no other way of explaining how the objects ended up on an ice floe off the southwest coast of Greenland in 1884.

Explorer that he was first and foremost, Nansen saw this polar current—now recognized as the Trans-Polar Drift—as a means that would enable him to reach the North Pole by sailing with the ice rather than against it. With that in mind, he and Otto Sverdrup worked with naval architect Colin Archer to design and build a boat that would be able to withstand the crushing forces of moving ice.

Although ice-worthy, the *Fram* never did make it to the North Pole, but the expedition was far from being a failure. During the three years that Nansen and his colleagues drifted with the ice, they made a number of discoveries, one of which proved that the polar basin was not shallow and island-riddled as some had assumed. It was, on the contrary, so deep that a 6,200-foot (1,900 m) hemp line was unable to touch the bottom. Only when Nansen and his crew tied ropes together were they able to get to depths that went to 13,100 feet (4,000 m).

On that voyage, Nansen and his colleagues also established many of the modern scientific principles that would serve oceanographers later on in the twentieth and twenty-first centuries. The most profound of these was the contention that "oceanographical conditions of the North Polar Basin have much influence upon the climate, and that it is equally evident that changes in its condition of circulation would greatly change the climatic conditions."

"There's not a heck of a lot of difference between what we're trying to do on this trip and what Nansen did more than a century ago," said Eddy Carmack. We were in the *Louis*'s conference room, and Carmack was poring over maps that showed all the station stops we would be making in the waters in between Baffin Island and Greenland over the next few days.

"Even our methods are the same," he continued. "It was Nansen and his buddy Helland-Hansen who came up with the idea that temperature, density and salinity were all you really needed to know to identify different kinds of seawater. The way we go about it now is maybe a little more sophisticated, but the principles are the same. There are a number of us who have spent a good part of our careers testing his theories and much of what he had to say back then is true now."

Carmack was not only the architect of this expedition, he was also the co-conspirator behind the historic North Pole journey of the *Louis* and the *Polar Sea* in 1994. Although tempting and appropriate to compare him to Nansen, Carmack is more like Edward Ricketts, author John Steinbeck's best friend. Ricketts was the inspiration for Doc, the charismatic intellectual who runs a lab along a stinky strip of sardine canneries and honky-tonks in Steinbeck's novel *Cannery Row*. Like Ricketts, the "father of fishboat science," Carmack spends as much of his spare time as he can sailing up inlets along the B.C. coast on a converted trawler. And just as Ricketts was fond of music, Carmack brought along his guitar in case there might be opportunity to jam with whomever else on board might be musically minded.

The slightly lazy, laid-back nature of his manner of speaking also made it clear that he wasn't originally from Canada. "Arizona," he said, correcting me when I guessed Montana. "I was newly hooked in scuba diving and that's why I got into oceanography. To find an ocean and to get out of Arizona."

The decision to pursue this course of study at the University of Washington was fortuitous because it was through that institution that he met Knut Aagaard, the American oceanographer who is related to Roald Amundsen on his father's side. Aagaard had just returned to the University of Washington from a postgraduate study of the Greenland Sea when the two first met in a two-man research shack drifting on the pack ice of the Beaufort Sea. The two hit it off almost immediately, first as student and teacher, then as friends and collaborators, and finally as co-conspirators in getting the Coast Guards of Canada and the United States involved in oceanographic research in the Arctic.

The Arctic was not where Carmack saw himself when he started his professional career at the Scripps Institute of Oceanography in California. It had been his intention early on to combine his passion for deep-sea diving with a research project in Antarctica. Still, he and his Canadian-born wife, Carole, dreamed of a life in Canada. There being no jobs available in oceanography north of the border, he went to work for Environment Canada instead, living in Vancouver and studying freshwater lakes in the interior of British Columbia and the Yukon.

Carmack has never second-guessed that move, even if it did take him away from the sea for a time. "That's how I met Andy and Carole Williams," he said, knowing I had just come from the Kluane Research Station in the Yukon. "I spent four years collecting data there looking at circulation in the lake. These were wonderful years where I kind of reattached to the beauty of high latitudes."

Carmack transferred to the Institute of Ocean Sciences in 1986, less than a decade after it got its humble start in an old seaplane hangar near the airport in Victoria. He joined a new project to study the environmental consequences of oil and gas development in the Beaufort Sea. Wonderful as the new job was, the timing couldn't have been worse. The price of oil collapsed and the oil companies packed their bags and went home. The Canadian government was also slashing budgets right across the board throughout the 1980s. Because science never had much of a voice at the political decision-making table, the cuts to research budgets were particularly merciless.

Marine scientists such as Carmack were also being blamed for failing to anticipate the collapse of the cod and salmon fisheries. "There were some dark times in the beginning," Carmack conceded. "Everyone looked at us like we should have had all the answers, but no one appreciated that we knew almost as little about the Atlantic and Pacific back then as we know about the Arctic Ocean now. And truth be told, no one was listening when some of us told them what was happening."

Still, that didn't stop Carmack from thinking up the big idea of sending a team of North America's best ocean scientists on a two-icebreaker journey to the North Pole in 1994. "The idea started as a joke," he explained in an almost apologetic way. "I was at a U.S. Polar Research Board meeting in 1989 when Sweden presented a plan to send their icebreaker—the *Oden*—to the North Pole through the Greenland Sea. While they were making the presentation, I turned to Knut Aagaard, who was there as well and said, 'Knut, you call your Coast Guard and I'll call mine, and we'll come up the other side and meet them at the North Pole.'"

The two had a good laugh. Reaching the Pole had been their shared dream for years. No one but the Europeans seemed to have the foresight or the proud history of exploration even to consider studying this unexplored part of the Arctic. Still, neither one could get it out of their heads when they met for drinks later than night. Like Nansen's plan to float the *Fram* to the North Pole, this seemed like a brilliant idea to further advance the study of the Arctic Ocean.

"It may sound bizarre now with so much news of Arctic ice melting, ice shelves collapsing and polar bears facing starvation," said Carmack. "But back in 1989 when the U.S and Canadian and even the European governments were being told over and over again by this climate panel and that climate committee that more information was needed to understand global warming, almost no one saw the Arctic Ocean as a missing piece to the puzzle."

Back at home, Carmack contacted Captain David Johns, the head of Arctic operations for the Canadian Coast Guard. He had no idea that Johns was a "big idea" guy like him. With plans for a Polar 8 icebreaker for Canada still in the works and the *Louis St. Laurent* in dry dock getting a multimillion-dollar makeover that would end up costing more than the price of a new ship, Johns had reasons and the opportunity to think big.

"My strategy was pretty simple really," said Carmack. "The story behind it is a much longer one than what I'm telling you. But basically, I asked him, very politely, whether I could borrow his biggest and most powerful icebreaker so that we could go to the North Pole. I cringed, waiting for him to slam the phone back on the hook, but he kind of paused and didn't say no, which was good. When he asked if there was a scientific purpose to this, I knew right then that we had a chance. In the end, it took five years of planning to get us there, but it worked. And thankfully, the

Coast Guard still answers the phone when I call. Otherwise, we wouldn't be here today."

"Prepare for chaos," said Bill Li when chief scientist John Nelson announced plans for a mock trial of everything we were going to do over the next several weeks. Li was a microbiologist from the Bedford Institute of Oceanography and an old hat when it came to deep-sea investigations. "This is the first time most of us have worked together," he explained. "It's also the first time that some of us have been on a big ship like this."

It was an eclectic group of scientists that sat around the table that day. In addition to Bill Li and Ed Hendryks, there was Connie Lovejoy, who had been part of an international team of scientists that had recently found new life forms in the Arctic Ocean. With justifiable fanfare, they dubbed this new group of microscopic organisms picobiliphytes; *pico* because they are measured in millionths of a meter, *bili* because they contain biliproteins, which are fluorescent substances that transform light into biomass, and *phyte,* out of recognition that they are plants.

Vlad Kostylev, another Bedford Institute scientist, wasn't expecting to find anything quite as astounding with the camera that he planned to shine on the ocean floor. He and Ed Hendryks, who would come to be known as the "muddy buddies" because of their interest in life on the Arctic Ocean floor, were looking for things that didn't require the use of an electron microscope.

"We're not going to get a lot of answers on this trip," Li said to me that day. "You fly over a forest one year and you don't expect to see a lot of changes when you come back the next. It's the same with the Arctic Ocean. What we see now is not going to tell us a lot about climate change in the short term, but it will be important in understanding what's happening when we look at it again in ten, twenty-five and fifty years. Hopefully, that's when a lot of the answers will come."

At that first meeting, I was assigned to a science team that included Bill Li and Sarah Zimmerman, chief scientist on a similar expedition to the Beaufort Sea the year before. We were to be on call from 4 p.m.

to 4 a.m. to draw seawater from a Rosette, a $500,000 carousel system that collects water in thermal bottles that are remotely tripped open at depths of up to 13,100 feet (4,000 m) or more. Much as I appreciated the opportunity, I didn't have a clue as to what I might be doing during these station stops, which the scientists called casts. Kindly, Li told me not to worry because it would become painfully simple in the next few days.

In the meantime, I had plenty of time to get myself adjusted to life aboard the ship. Fortunately, I didn't suffer from the seasickness that plagued Charles Darwin nearly every day he was aboard the *Beagle* when it sailed around the world. "It's no trifling evil," he wrote to a friend after the boat rounded the notoriously rough waters off the Cape of Good Horn. Darwin suffered so much on that journey that he attributed it to his perennial ill health later on in life. He wasn't alone. John Franklin and Lord Nelson were sailors who were just as afflicted.

Sleep, on the other hand, did not come easy. Not only did the night-shift hours throw off my biological clock, the cramped, airless quarters of my tiny windowless room belowdecks was too claustrophobic for me to get comfortable and relaxed. Metal doors clanged constantly. The engines moaned and throbbed. A strange hissing sound periodically came though the ventilation system. All I could do was turn the oscillating fan on high to try to drown it all out.

The first serious test of my seaworthiness came on the night we left the Strait of Belle Isle and headed into a gale on the Labrador Sea. The winds rocked the boat so hard that night, I woke up from a rare moment of sleep, clutching the frame of my narrow bed. It felt like we were tobogganing down a hillside in a cardboard box.

Unable to get back to sleep, I got up and had coffee on the portside, wind-sheltered deck of the ship with Bryon Gibbons, the first officer. He was up at 4:30 a.m. every day. "Get used to it," he told me while we did our best not to spill coffee from our mugs. "Icebreakers are designed to have smooth bottoms, so they roll more than your traditional ship. You put this boat on flat ground under a forest of trees in a national park, and if there was dew on the ground in the morning, she'd rock and roll."

The *Louis* has two dining rooms, one with windows, waiters, white cloth tables and a cappuccino machine that I discovered too late in the trip to do me any good. This room was reserved for the officers and the senior scientists. The other was a cafeteria-style galley for the rest of us. Food there was good but greasy and stacked far too high on the plates. Worse still were the pies and cakes that were available all night long. The coffee, however, was so weak, I could see the stem of my spoon halfway down when I stirred in sugar, which I usually never use except on this trip, when I needed an extra kick to keep me going.

John Wells was tucking into an enormous plate of eggs, sausages and home fries when I found him in the galley eating alone one foggy day at around 6 a.m. I had just got off a shift that took longer than normal because we were sampling 1,000 feet (300 m) deeper than we had in previous days. The deeper we went, the longer it took to get the Rosette down to those depths.

"My goal is to see a polar bear," the Newfoundlander confided. "Before I die, I want to see a polar bear. Preferably, I'd like to see this polar bear with two cubs. And it would be good if this bear were killing a seal at the time. But I'll take what I can get. A polar bear walking on the ice toward the ship will do just fine."

In his other life, Wells is an associate in the Physics Department at Memorial University in Newfoundland. He was here on contract with the Canadian Wildlife Service. His one and only responsibility was to count and jot down the name of every bird and marine mammal we saw along the way. Whenever the ship was moving in daylight, he was on the upper deck with binoculars glued to his eyes. The only time he took a break was when he slept or ate or when the ship was stopped to take samples. There was no one there to cover for him. He ate fast and slept periodically. If we were to meet up with a polar bear, he'd likely be the first to see it.

By this point in the trip, we had already seen a pod of killer whales, several pilot whales, dolphins and countless seals swimming alongside or within eyesight of the ship. Heading into the Davis Strait from the Labrador Sea, we left the ducks and seabirds behind. Thick-billed murres, dovekies and fulmars, said Wells, were about all the birds we would see from here on in.

This was Wells's fifth voyage on an icebreaker. "The first time, there were no women," he complained. "The second time, there were no women. The third time," he stated more emphatically, "there were no women. The fourth time, there was one. This time, they seem to be everywhere. It's wonderful."

This was true. Thirty years ago, there were virtually no women on voyages like this one. Back then, it was not uncommon for an icebreaker to be out for one hundred and twenty days or more without a crew change. Now, a shift lasts no longer than three weeks and women fill virtually every role there is to fill on an icebreaker. Of the more than forty principal investigators associated with the Canada Ocean Monitoring Experiment, about one-third were women. Romance and rumors abound. Most of the relationships are short-lived, but some become long-term affairs or lead to weddings. Inevitably, there's gossip when a couple tries to be discreet. Big as the *Louis* is, it's almost impossible to keep things secret among groups that are so closely knit.

Later that morning, the CBS television crew on board got into hot water for putting time-lapse cameras up in the science laboratory without official permission. Daniel Sieberg, the science reporter for the American network, was on the ship partly because climate change had become such a hot subject in the United States and partly because his father, Doug Sieberg, the expedition's chief technician, had suggested the idea for a series of stories.

In some ways, Doug was closer to the *Louis* than anyone else on board. He was the man who everyone went to whenever there was a problem with the scientific gear. That's why he never slept. Things were breaking down at all hours of the day. Doug was also one of those who fell in love on a voyage. He married his second wife on the *Louis St. Laurent* when the ship sailed to the North Pole in 1994. Carmack was his best man. Although he officially retired in 2005 after thirty-one years with the Institute of Ocean Sciences, Doug continues to come along for the ride.

"It's an addiction, this Arctic world," he told me. "And I'm brute for punishment. Every time I think this is the last trip, Eddy calls, and I find myself unable to say no. My problem is that I was born two hundred years too late, I would have loved to be an explorer on one of those early voyages to the Arctic."

I had nothing to do with what the CBS crew had been up to, innocent as it was. Still, I got called into the captain's office to hear the same speech he delivered to them minutes earlier. Captain Andrew McNeill, I discovered in fairly short order, was not someone who was going to open up and share a lot of insights and experiences with me. He had a no-nonsense demeanor about him that reminded me of a high-school science teacher who sees in every student the potential for a lab to be blown up. And who was I to blame him? The man had twenty-eight years' experience on Coast Guard vessels. He was involved in the 1985 Air India Salvage Operation, which followed the explosion that killed all 331 people on board the plane. He was on the deepest cable repair ever undertaken by the Coast Guard and worked with the RCMP to make arrests on the high seas. With at least one employee on board who was out on prison leave, and young scientists falling over one another everywhere, he didn't need a journalist like me doing as I pleased. I left the room feeling like a kid who'd been punished for something he hadn't done.

John Wells was sympathetic when we met later that night for supper. He had just been rebuked by the first officer for whistling on deck. Whistling aboard a ship, he knew, is frowned upon because it is thought to bring ill wind. Much as Wells tried, he couldn't get the tune, the theme music for the old television show *Dallas*, out of his head. This confession led to a heated discussion about who shot J.R., the villainous character on the show. No one at our table could remember. But Wells observed that Bryon Gibbons, the first officer, was a dead ringer for Joe Mannix, the mustachioed detective character in another popular television show.

The more time passed, the loopier these discussions at the dinner table became. Part of it had to do with the isolation. With nowhere to go but the galley at breakfast, lunch and supper time and the lounge for drinks between 8 p.m. and 10 p.m. (if you weren't on call, as I almost always was), some people got a little cagey. Others were simply giddy from the long hours and the repetitive nature of the work.

Romantic as the field of oceanography might seem, I quickly learned, as Bill Li had warned me, that there's nothing sexy about milking icy water out of twenty-four Niskin bottles attached to a Rosette between 10 p.m. and midnight and then again from 2 a.m. and 4 a.m. as we did three days in a row while heading into the ice in Davis Strait. In an effort to stay awake and maintain sanity in this cold and often fog-shrouded world, those in charge of the music would play tunes that would have been even more maddening under normal circumstances. Particularly popular was Boney M.'s hit, "Rasputin." The tune was, embarrassingly, infectious, and in its own weird way, it was wonderfully diverting watching two serious scientists such as Sarah Zimmerman and marine biologist Diana Varela wearing hard hats, lab gloves and steel-toed rubber boots dancing to the disco beat.

Yet there was always a crisis that would bring everyone back to his or her senses. One day it was a half-dozen Niskin bottles imploding under intense pressure at 3,000 feet (900 m). Another day, it was oxygen bubbles that were traced back to the samples our team collected. Since the only way of removing those unwanted bubbles is with a syringe, we had to drop our glacial pace even further to ensure quality control. This extended the midnight shift by an hour or two for a few days.

Diana Varela, the oceanographer from the University of Victoria, suffered the most. She was focusing on phytoplankton, which are an important part of the climate change puzzle because these tiny plants form the base of the food web and trap carbon when they die and sink to the ocean floor. There are all kinds of phytoplankton in the Arctic Ocean. The largest look like flecks floating in the water, but most appear as brown- or green-colored water and can only be identified with a microscope.

One of Varela's goals was to find out how phytoplankton use nitrogen, silicon and carbon, the three elements that dictate their rate of growth in the ocean. In order to do so, she needed to keep her samples cold. Unfortunately for her, there was no pump available to draw cold seawater to the ship's deck. So when the skies cleared and the temperatures rose, as they did for a few days, she and a handful of volunteers were forced to haul dozens of buckets of ice up several flights of stairs to keep their storage tanks cool.

John Nelson, the chief scientist, was well aware that everyone was overworked and overwhelmed. But neither he nor Eddy Carmack was interested in doing anything that might slow the ship down. "Sometimes like a horse, you've got to pull your ears back and run as fast as you can," Carmack said on the day the issue of overwork was raised at a meeting. "We've got to make the most of this opportunity. The satellite maps show that there is a lot of ice ahead of us that's going to slow us down. And we're behind already."

The late-evening sun was crimson red and shooting orange flares over the mountainous coastline of Baffin Island. Here the ice "cracked and growled, and roared and howled," just as Coleridge wrote in "The Rime of the Ancient Mariner." In the ship's wake, the brash ice sparkled in the fading light. The sun was now setting for only a few hours in the very early hours of the morning.

Even though it was late, the working day was not over. On the upper deck on the bridge, Captain McNeill opened a window and casually tossed a bottle of Sleeman Honey Ale into the ocean. Down on the foredeck below, some twenty scientists, red-faced from too much sun and windburn, let out a collective cheer and then did the same thing. Much as some might have hoped, there had been no beer in the bottles. Nor were we littering. Each one of the bottles contained a message, which requested the finder record the site of discovery and the drop number in the bottle. That number, once retrieved, could be checked against a list of drop locations recorded in a computer that night.

Oceanographers have been using drift bottles as a means of studying ocean currents since AD 310, when Theophrastus tried to prove that the waters of the Mediterranean came from the Atlantic. There is no evidence to indicate whether Theophrastus received any response. Carmack has had much better success since he launched the idea in the Arctic in 2000. Since then, about 4 percent of the bottles have been retrieved, usually after about a two-year voyage. The rest have sunk, leaked so that no one could read the note inside, or washed up on remote shores where there is not much likelihood of someone finding them.

Although crude, this experiment is an effective way of testing the theory that runoff from the continents drives clockwise coastal currents around both North America and Eurasia. It also demonstrates how relatively fresh and cold, and warm and salty water moving in and out of the Arctic Ocean can interact with the air and sea ice in ways that have a profound impact on both marine life and the climate.

This happens in more ways than Matthew Fontaine Maury, August Petermann or Fridtjof Nansen could have possibly imagined. Various studies, including those by Humfrey Melling, leader of thirty oceanographic expeditions to the Arctic, have recently led to new insights into the processes which control the flow of water and formation of ice in the Arctic Ocean. Much is still to be learned, but what we do know from those studies is that the main flow of Arctic water is from west to east. The inflow begins with relatively fresh water from the Pacific entering the Arctic through the Bering Sea. The inflow is cold in winter and warm in summer. In all seasons, it is extremely nutrient-rich and the main reason why the Bering and Chukchi seas are two of the most biologically productive oceans in the world.

When the frigid winter winds blowing off the coast of Alaska freeze this water and send the ice out to sea, the salt expelled from the ice dissolves into the water left behind. This heavier, salt-laden water eventually sinks and spills over the continental shelf into the Canadian Basin. When it comes into contact with an Atlantic current carrying warm, much saltier water counterclockwise around the Arctic from Fram Strait in between Greenland and Spitsbergen, the lighter water from the west naturally settles on top. As a result, the warm current trapped below can only lose its heat to the overlaying blanket of icy cold water. None reaches the ice at the surface. Without the cold blanket, says Melling, there would be not nearly so much ice in the Arctic.

In the Beaufort Sea, strong winds send this inflow of relatively fresh Pacific water spinning into a huge gyre that circulates in clockwise fashion over a 1.2-million-square-mile (2 million sq. km) area. When these winds weaken, as they do every ten years or so, large volumes of this water circulating in the Beaufort Gyre leak out, passing through several gateways between the islands of the Arctic Archipelago before spilling out into two main channels leading into the North Atlantic.

Carmack, who collaborates with Melling at the Institute of Ocean Sciences, compares this circulation of surface water to an air conditioner that keeps the northern hemisphere cool. Any disruption in the system that drives the air conditioner, he says, not only has the potential to lead to a change in the temperature, salinity chemistry and ultimately the marine life in the ocean, it can also affect the climate on a global scale. Like El Niño's warm currents in the Pacific, changes in Arctic Ocean circulation can alter the atmosphere in ways that might exacerbate droughts in drought-stricken areas and strengthen hurricanes in those areas prone to storms.

Decade-long variations in Arctic Ocean circulation that we know have occurred regularly at least since around 1970, however, might be influencing the climate in ways that are independent of the global warming generated by human activities. A NASA team led by James Morison of the University of Washington found this out when they used Earth-observing satellites and deep-sea gauges to measure changes in the weight of water in the Arctic Ocean column between 2002 and 2006. Data from those monitors showed that the ocean circulation changed from the counterclockwise pattern that was dominant in the 1990s back to the clockwise pattern that was dominant before that time. They attributed this reversal to a weakening of the Arctic Oscillation, a recurrent pattern of atmospheric circulation in the northern hemisphere. As a result of this weakening, the surface water of the ocean became less salty and lighter.

Morison and his colleagues were reluctant to link the recent decadal-scale changes in Arctic Ocean circulation to global warming when they reported their findings. But most climate models predict the Arctic Oscillation will strengthen its counterclockwise pattern in the future. Even Morison suspects that the events of the 1990s may have been a preview of how the Arctic will respond over longer periods of time in a warming world.

Sometime around 1990, an atmospheric low pressure began to nudge out the massive highs that traditionally dominate winters in the western Arctic. Low-pressure systems brought wind, rain or snow and generally warmer temperatures to the North. The Arctic Ocean responded to this shift in the atmospheric regime by allowing a larger amount of warm, salty water from the North Atlantic to gush in from the east. At the same time, scientists believe that cold Pacific water may have begun leaking out through the Arctic Archipelago into Baffin Bay, Davis Strait and the Labrador Sea. This occurred right around the time the cod fishery was collapsing.

No one in the Department of Fisheries and Oceans paid much attention when Carmack and his colleagues Fiona McLaughlin and Robie Macdonald sounded the alarm, warning that the cold water may undermine the recovery of a cod fishery that was in a state of collapse as a result of overfishing. Carmack still uses this example to illustrate how Canada's three oceans are connected in a very fundamental way and must be studied as such.

The ecological shifts in the Arctic Ocean are continuing. Due to warmer temperatures and a weakening of cold winds blowing off the coast of Alaska, the ice in the Bering Sea has been melting sooner. In those waters of the Bering Sea that are now warmer, Carmack, American marine biologist Jackie Grebmeier and others have seen the production of phytoplankton that traditionally grow in cooler, icier conditions decline dramatically. With fewer phytoplankton for worms, krill, shrimp, clams and amphipods to feed on, bigger bottom-feeding species such as walrus and gray whales are being forced to go farther north to find food. Some cold-water fish species such as pollock and pink salmon are doing the same thing.

Having to travel farther for food is especially hard on gray whales. Their migration from Baja California in Mexico to the Bering Sea in late summer and fall is already the longest marine mammal journey in the world. Adding more miles takes more energy. It is also unsustainable. As the Arctic waters become more sub-Arctic in nature, Grebmeier points out, these whales will eventually have smaller and smaller pockets of habitat to move into.

Shifts in the population of gray whales might, surprisingly, have an impact on phytoplankton. It's long been thought that phytoplankton are the foundation on which larger Arctic life forms depend, and this is true in many ways. Phytoplankton are taken up by the amphipods and copepods that feed the arctic cod. In turn, the arctic cod plays a key role in the diet of seals, many whales, seabirds and other larger fish. But growing evidence suggests that the relationship between those at the bottom of the food chain and those at the top is more of a two-way street than biologists originally thought. Bill Li was explaining this to me one very long night when we were waiting for the Rosette to come back to the surface for a second time. Remove a larger predator such as a polar bear, a gray whale or narwhal from the equation, he noted, and you may very well alter the nature and makeup of the phytoplankton in the water column.

"That's how closely intertwined and fragile this marine world is," he said "That's why I keep saying that looking at just one piece of the puzzle over a very short period of time is not going to give us the big picture we're trying to put together."

Up on the foredeck, Ed Hendryks and Vlad Kostylev, "the muddy buddies," didn't need a microscope to put together the pieces of the puzzle they were looking for. As Hendryks pulled brittle stars, brachiopods and giant bamboo worms out of the ocean mud, through his camera Kostylev was spotting red fish, flat fish, sea squirts, anemones and other sea creatures hovering above them.

The picture detail was so fine in some cases that we could see the eyes of one of these tiny creatures shifting back and forth, trailing the movement of the camera. But the find that really put a smile on Vlad's face was the coral he identified at the bottom of Davis Strait. "This is so cool," he said with his thick Ukrainian accent. "I've seen this off the coast of Nova Scotia, but I would never have expected to find it so far north."

Surprising as the find was, it is not unusual, considering Connie Lovejoy's discovery of picobiliphytes and the fact that Archaea, those tiny single-celled extremophiles that have no nucleus and no mitochondria and that are unlike any other life form on Earth except perhaps bacteria, were only discovered in the ocean a few decades ago. Now they represent a third domain in the Tree of Life model that Darwin used, and microbiologist Carl Woese later refined, to visually explain the interrelatedness of all living things that evolved from one common ancestor. Eukarya, Bacteria, and now Archaea represent those three domains.

Bacteria are commonly associated with human and animal diseases. The Earth is teeming with countless numbers of species, many of which play an important role in nitrogen fixing and decomposition of organics. Eukarya are those organisms whose cells contain a nucleus. They include everything from plankton and fungi to animals and humans. Archaea are in a world of their own. They are the methane-makers that live in swamps and marshes and in the deep sea; the salt-loving halophiles that live in the Dead Sea and Great Salt Lake and the thermophiles that live in extremely hot or extremely cold environments. The recent discovery of volcano activity on the Arctic Ocean along the so-called Gakkel Ridge holds promise for the discovery of more of these life forms.

Carmack has no doubt that more weird and wonderful things are going to be discovered in the Arctic Ocean. That's another reason why he believes there is currency in better understanding what is happening in this underwater world. "In many ways, it's a black hole that we're looking into," he told me that night. "The Arctic Ocean is among the least explored wildernesses in the world. To have such an enormous region be so poorly understood exemplified the three gaps that challenge the leadership of our nation. The first is the 'knowledge' gap, or what we don't understand. This is the job of the scientist. The second is the 'science-policy' gap, or how to get new information into the policy realm; this is the job of bureaucrats. The third is the 'action' gap, or how to get knowledge working for us all to make Earth a sustainable home for humans; this is the job of politics."

We finally finished the last cast off the north coast of Baffin Island at 2:30 a.m. The ship was now heading into Lancaster Sound, the legendary entrance to the Northwest Passage. As usual, sleep didn't come easy.

Here we were grinding through land-fast ice, which unlike the soupy grease ice, patchy pancake ice and first-year ice we had already passed though, requires a little more engine power. The pounding and grinding that came each time we hit a thicker chunk of ice made it feel as if I was locked in an empty 45-gallon drum that was thrumming against rocks on the seashore.

Sometime around 6 a.m. a big thud nearly knocked me out of bed. This was followed by a series of lighter knocks. In a fog, I was confused but sensible enough to go to the door to see if it might be an emergency. I found Bryon Gibbons, just about to walk away.

"I'd been knocking for a good minute," he said. "Thought you'd like to see a polar bear having breakfast on the sea ice."

I scrambled to find my camera and clothes. Even though I was still half asleep and unable to get my eyes to focus, I managed to locate the stairs to the foredeck. Up there, people in various states of undress were standing in the bitter cold watching silently as the ship passed a polar bear and her cub feeding on a seal that they had just killed on the ice. I had seen hundreds of polar bears in the wild, but never anything quite like these two. Coated in blood, they paid us no notice until the ship passed within a few hundred yards.

It didn't occur to me at the time that John Wells was not there to see them. I expected that he would have been on the bridge as he always is when the ship is moving. When I found him in the galley a short time later, he was mortified. "My Lord," he said. "The day my alarm clock doesn't go off, the ship nearly hits a polar bear and her cub. I'm cursed."

It didn't take long for Wells's luck to turn. Later that afternoon, a polar bear was spotted swimming in the water on the starboard side of the ship. By the time I got up to the bridge to make sure that Wells was there to see it, the bear was already hopping from one ice floe to the other, trying to make as much distance between him and the approaching ship. Normal as a scene like this would be at this time of year, it still struck me as melancholy. The coming years would provide similar watery scenes, but the sea ice would be melting sooner with each season. Like the gray whales, which are being forced to move into smaller and smaller pockets in the Bering Sea, these bears may soon run out of room and lose the platform on which they hunt, mate and rear their young.

In the late summer of 1818, British explorer John Ross was in command of two ships, the *Isabella* and the *Alexander,* when they sailed into Lancaster Sound in search of a northwest passage to the Far East. With "no appearance of a current, no driftwood and no swell from the northwest," it was the opinion of Captain Edward Sabine that there was "no indication of a passage." The next day, Ross sailed west to have a closer look, leaving the *Alexander* behind. Within the hour, he was slowed by fog and eventually stopped by what he described as a "high ridge of mountains, extending directly across the bottom of the inlet." Seeing that there was a sheet of ice that lay in front of it, Ross turned back, certain that there was no passage in this direction.

The expedition wasn't entirely a failure. Ross went home with reports of several new species, including muskoxen, arctic fox, arctic hare, Sabine's gull and brittlestars collected from the ocean bottom. But Ross was never forgiven by John Barrow, the head of the British Admiralty, for failing to go farther and discovering what other explorers would later learn, that no mountains blocked this route and that a northwest passage to the Pacific began here.

Nevertheless, the explorers who later sailed through Lancaster Sound were stopped time and again by the thick multi-year ice that pours out of Viscount Melville Sound and the M'Clintock Channel to the northwest. Alexander Armstrong, the surgeon aboard the *Investigator*, one of several ships that went in search of the missing men, described the futility of any attempt to get through the ice that encased the Arctic world:

> No earthly means of magnitude or power, aided by all the best appliances of art, and guided by the judgment, ingenuity, and best energies of man, could avail in the slightest degree, in surmounting the overwhelming obstacles, which, on these occasions, opposed our progress.

More than one hundred and fifty years after Armstrong described the impassability of that Arctic ice, the *Louis St. Laurent* was steaming through Lancaster Sound on route to the last cast on this first leg of the trip. It was a gorgeous evening. The skies were clear, the winds were calm and the air was soft and warm for the first time in more than a week. An air of giddiness prevailed. That day we had seen four bears, a walrus and a flock of rare ivory gulls that appeared, somewhat magically, in the foaming waters left in the wake of our passage.

All night long, the station stop kept getting pushed back. No one seemed to know why. Sometime around midnight, I realized that I wasn't going to be able to last much longer. The cold, the fog, the incessant ice and the impossibly long days had finally gotten to me after nearly three weeks of sailing. The enervating heat from the sun was now overwhelming. I asked Bill Li if he'd give me a wake-up call when the Rosette was raised to the surface. The next thing I knew, it was 6 a.m. I was in a panic, thinking that I had somehow slept through the knock on the door and let my side down. Bill Li just laughed it off when I caught up with him and apologized. "The Rosette didn't come up until exactly 4:30 a.m.," he said. "We didn't have the heart to wake up the next shift. Nor you. This being the last one, it was no problem."

Later that day, I was asked if I was interested in going on a helicopter flight with the Canadian Ice Service to check on the conditions that lay ahead. Normally, a flight like this is designed to help ships and smaller vessels pick their way through thick ice in a way that a satellite image cannot. After fifteen minutes in the air, there wasn't much to see, only an endless stretch of emerald green water that shimmered with specks of white. At this point, the pilot turned his back on the water and headed toward the Croker Glacier on Devon Island, close to where John Ross saw his mythical mountains. Here, a series of mostly snowless valleys stream out of a giant ice cap, which sits on a plateau atop the island. This brown, arid world looks so much like Mars that NASA has been sending scientists here since 1999 to help them understand how life began on the Red Planet.

A half-mile (1 km) up from the toe of the Croker Glacier, the pilot put us down on one of the few spots that was not streaming with meltwater. Satisfied that we were stable, he signaled us to get out. Here, as it was on the Brintnell and the Lowell glaciers in the Yukon, the landscape was too big and too alien for the imagination to grasp it quickly. A few feet above 50 feet (1 m above 15 m) of ice, one would expect to be in a deep freeze. But with no wind and so much of the sun's heat reflecting off the glacier, I was forced to strip down to my T-shirt. In deep stillness, and the blazing heat, I felt a mystical summoning, similar, I thought, to those that haunted the explorers who first passed through this world. Even though it had been explained to me by scientists on board the ship, I still found it remarkable that a glacier, so large and lifeless, can be so light and transforming when it melts into rills of freshwater that trickle into the salty sea.

Knowing that this glacier would one day turn into one of those arid valleys on the island, I bent down, cupped my hands and took a ceremonial farewell drink. Armstrong, I realized, was wrong. There was an "earthly means of magnitude or power" aided "by the energies of man" that was capable of surmounting something as formidable as this.

chapter five

ARCTIC OUTBREAK

– Repulse Bay, Nunavut –

Inuit are being poisoned from afar by toxins—PCBs, DDT and other chemicals—carried to the Arctic on air currents. These chemicals contaminate the food web we depend upon: seals, whales, walruses and end up in our bodies and the nursing milk of our mothers in high levels. So what a world we have created when Inuit women have to think twice about nursing their babies.

—Sheila Watt–Cloutier, Inuit Circumpolar Conference 2005

East Bay is the name of a tiny, uncelebrated national bird sanctuary off the east coast of Southampton Island at the north end of Hudson Bay. The island is flat, treeless and about one-third the size of New York's Central Park. Although small and bleak, it is a spring and summer home to a colony of common eiders, the largest and quite possibly the most social of all ducks in North America. During the egg-laying season, when nesting females go for three weeks without feeding, so-called "aunts" will gather round and lead the newborn ducklings to water. To keep themselves warm in the coldest, darkest nights of winter, they stop eating and moving to conserve energy. They huddle into tightly knit groups so that one duck blends into another in such a seamless fashion that they can't be counted.

Canadian Wildlife Service scientist Grant Gilchrist and his colleagues were on the island in the summer of 2006, continuing their long-term study of the five thousand to ten thousand eiders that traditionally nest here. Throughout June and most of July, the tagging operation went as smoothly as could be expected in an area that is often shrouded in fog and hammered by the occasional summer blizzard. The only big excitement came from the polar bears that sometimes wandered into camp, and they, too, were to be expected.

The routine nature of the days changed in a dramatic way when dozens of birds on the island went into convulsions. Many of them flopped about before collapsing. Others simply died sitting on their nests with their eyes still open. Gilchrist and his colleagues were helpless to do anything. In just hours, the dozens quickly became hundreds and then thousands. Even before rigor mortis set in, predatory gulls flew in and pecked away at the dead and the dying. In fewer than two days, it was all over. More than thirty-five hundred birds had died; all of them females because male eiders fly off once the nests are set up. That summer, three out of every four nesting females in the colony had been removed from the population.

The avian cholera that killed these birds was first detected in North America in the winter of 1943–44, when the disease struck waterfowl overwintering in the Texas Panhandle. The bacterium has been steadily moving north ever since, reaching the Gulf of St. Lawrence in 1964 and the north end of Banks Island in the High Arctic by 1995. Thirty thousand snow geese died on Banks Island in 1995. Another twenty thousand died a year later. Outbreaks of avian cholera are so common now that more than one hundred bird species on the continent have been infected.

With just the few exceptions, the Arctic has been largely spared the ravages of this and many other diseases and parasites that afflict wildlife populations farther south, but that summer marked the third time that avian cholera has hit East Bay since 2004. It was evident to Gilchrist, as has it had been to disease specialists for some time, that the Arctic may no longer be immune to the new and emerging diseases such as West Nile and Chronic Wasting Disease.

The spread of disease throughout the Arctic world is either the scariest threat that climate change will bring or one that may not matter all that much. At best, natural immunity will protect most species from diseases that have, up until now, been kept in check by brutally cold winter temperatures and by other climatic conditions that prevent disease-carrying animals from expanding their range northward. Some birds and animals that do get infected by new pathogens may tolerate the symptoms relatively well.

At worst a warmer Arctic could expand the range of disease-carrying species and set the stage for the kind of large-scale die-offs that have occurred at East Bay, Banks Island and in other parts of the world. In U.S. waters alone over the past twenty years, there have been fifty-seven significant die-offs of marine mammals. Many of those have been linked to diseases nurtured by changing climatic conditions. Included among them are the deaths of more than five hundred Pacific gray whales between 1999 and 2001 and the loss of thousands of Pacific Coast California sea lions between 2000 and 2005. The potential for any number of diseases and parasites to strike is as serious in the Arctic Ocean as it is on the tundra and in the mountains, where lungworm and sheep muscle worm are beginning to take their toll on muskoxen and wild mountain sheep.

It was early August when we were heading north along the coast of Repulse Bay, about 200 miles (320 km) northwest of where Gilchrist and his colleagues were once again camped, bracing for a fourth outbreak of avian cholera. Ours was a large group, fifteen people in three boats. Eight were Inuit hunters who had come along to assist four scientists, one graduate student and one Italian chef. These scientists were interested in what was happening to narwhal that come to feed in these waters in summer. Where they go, how deep they dive, how healthy they are and how they respond to ice and changing environmental conditions is a mystery that is only beginning to be unraveled.

The leader of our group was Jack Orr. His original plan was for us to spend a day traveling by boat from Repulse Bay to Lyon Inlet, a long, narrow stretch of water favored by narwhal. He and his brother Jim and many of the same people in this team had caught and released five narwhal there the year before after taking blood and tissue samples and putting satellite transmitters on the dorsal ridge along the narwhals' backs. It went so quickly and smoothly that Orr thought he'd go there again.

That plan, however, had to be canceled when satellite images and radio reports from Inuit narwhal hunters suggested we would likely be stopped by a massive sheet of thick ice moving south from Foxe Basin. Trying to get through that ice might not only prove to be futile, it could also be disastrous. Once ice locks a boat onto shore, it can be a long time before it sets it free—if it sets it free without doing serious damage.

The rocky spit of land on which we ended up camping was sheltered by a tiny island in Repulse Bay that stretched about 1 mile (2 km) from north to south. The only things that lived on this big hump of rock were a few gulls and some Inuit sled dogs that had been left there for the summer to fend for themselves. At the tip of this spit on the mainland was a small, rocky bluff that offered a panoramic view of everything that might be moving our way from the north. Laurent Kringayark, the Inuit captain of the expedition, felt it was as good a place as any to drop the nets and wait for the narwhal to come in. He and the other Inuit hunters had seen the whales pass through this narrow passageway many times before. Jack left it to the Inuit to decide. He believed that they knew better than him where the whales would be found and how to go about doing it. He saw them more as collaborators than hired hands.

Lying on the edge of the Arctic Circle, Repulse Bay would seem like the last place in the world where a newly introduced disease could take hold in a marine mammal population. There are just two narrow channels leading into it, Roes Welcome Sound from the northwest end of Hudson Bay and Frozen Strait, which connects to Foxe Basin. Apart from the sea lift that brings in food and supplies in August, what little marine traffic there is, is the kind that takes Inuit hunters to and from their seasonal camps. There are no mines, no oil or gas rigs within 600 miles (1,000 km) of these waters.

Despite the site's attractiveness, Jack Orr was game but not entirely happy with it. The annual narwhal hunt in Hudson Bay and Foxe Basin was still under way and with the community of Repulse Bay being just 9 miles (15 km) to the southwest of us, he was concerned that the narwhal might be hypersensitive with so many gunshots hitting the water. Fifty-three narwhal had been harvested so far, and the Inuit were having trouble getting the last seven they were allowed to take as part of their annual quota, set by themselves and government scientists.

Of lesser concern was a report of a pod of killer whales that had been spotted in the area recently. Narwhal would just as soon face the blazing guns of hunters in shallow waters than orcas in the deep. Out there, the big whales pulverize the smaller narwhal with a series of lethal blows that come from all sides. Attacks such as this used to be extremely rare in the Arctic. But now that much less ice chokes up the entry points into Hudson Bay and Repulse Bay in summer than in the past, the bigger whales have nothing to stop them from moving in. Whenever they show up, the narwhal flee to safer havens. Several years earlier at another narwhal site near Arctic Bay, Jack had witnessed what happens when orcas attack. All that was left of the narwhal that they had been watching offshore was a slick of oil that rose to the surface.

Whatever worries Orr may have had were relieved that first day. While we were setting up our tents, the radio tower and other equipment, a dozen narwhal swam past, seeming to take no notice of all the human activity on shore. "Let's just hope that those weren't the last of them," said Orr. "We had such good luck catching whales last year at Lyon Inlet, I was kind of wondering whether we would be as lucky this time around."

Orr had already smoked his first pack of cigarettes and was now halfway through the second, knowing that he was eventually going to run out of his rations if the trip took the full ten days. Each year, he goes home and finds a way of weaning himself off the noxious weed. And each year he comes back, only to get hooked again; the pressures associated with organizing an expedition such as this and waiting days for the whales to come, if they come at all, can be nerve-racking.

Fifty years old, thin, white-haired and fair-skinned, Orr is somewhat weathered by all the years he's spent in the Arctic. He got his start with the Department of Fisheries and Oceans in 1981 working on arctic char in Hay River and Cambridge Bay in the western Arctic. He also worked with University of Guelph scientist David St. Aubin in Igloolik, just north of Repulse Bay, where he first glimpsed narwhal. A pod got stranded in a pool of water that had become iced in from all but one side. Fearful of exiting over a shallow reef when the tide was coming out, the narwhal refused to budge. St. Aubin and Orr were called in to help drive them out to safer waters.

When the opportunity arose the next year to shift full-time to the marine mammal side of the Department of Fisheries and Oceans, Orr seized it, despite not knowing where exactly it would take him. The timing couldn't have been better. Tom Smith, a senior scientist in the department, was toying with the idea of attaching satellite transmitters to beluga whales in the Arctic. His aim was to find out where, when and how deep the mammals were diving in summer and where they were going when they left the Canadian Arctic waters in fall.

These early captures were based on harvesting techniques of nineteenth-century whalers, who used their boats to drive hundreds, sometimes thousands of whales into shallow waters where they could be easily shot or bludgeoned to death. During those first years, Smith, Aubin and others would put three or four men in one boat to corral the whales in a similar way. Once a whale was cornered in a shallow area, one man would jump out and put a lasso around the animal when it popped its head out of the water. If he was successful, another would follow with a soft rope that he tied around the whale's tail. Through trial and error, biologists improved the odds of getting the whales by using hoop, seine or stationary nets.

This, of course, would sometimes turn into a bit of a rodeo. Neither belugas nor narwhal are particularly aggressive once they are restrained. But there have been a few occasions when the big males have rammed a boat, charged at people in the water or bitten a leg. The first time I was involved in a capture at Cunningham Inlet off the north coast of Somerset Island with Tom Smith, a male turned on us during the chase and tried to take a bite out of the Zodiac. When Smith signaled for me to jump in, I made sure that he went first.

Efforts to catch narwhal came several years later. Unlike beluga, narwhal tend to avoid shallow waters where the seine net technique is most effective. The male's long tusk also makes a hoop net impractical. The only option then is to set a stationary net in deep water near shore and hope that one or more whales swimming past will get careless and run into it. Once the whales are in the net, three or four people will hop into the Zodiac and bring the animals back to a soft landing on shore.

Crude as some of these early capture efforts sometimes were, they and the more refined efforts that followed laid the foundation for the understanding of where whales migrate to, why some beach themselves and how viruses and bacteria take root in animals that are injured or immune compromised. When St. Aubin left Canada for the Mystic Aquarium in the United States and Smith moved on to pursue other challenges outside of government, Orr was left to lead the beluga and narwhal capture efforts in Canada. So far he has captured and tagged more than two hundred beluga whales and more than fifty narwhal.

Disease was the last thing on most scientists' minds in those early years. Certainly no one was contemplating the possibility that morbillivirus, a highly contagious group of viruses that can literally wipe out thousands of animals, if not entire populations, at a time, was present anywhere in the world's oceans. The measles virus that affects humans, the canine distemper virus that kills dogs and the rinderpest virus that affects cattle—these were the morbilliviruses that microbiologists and health officials were most worried about.

All that changed in 1988, when thousands of seals washed up on the shores of northern Europe and Siberia's Lake Baikal. Postmortems showed that the animals died of pneumonia brought on by canine distemper. No one knew where or how the virus got into the ocean. But when several more epidemics, and two more morbilliviruses—phocine distemper that affects seals and other pinnipeds and cetacean morbillivirus that kills dolphins and porpoises—struck in other places, microbiologists knew a very serious problem was emerging.

Still, there wasn't any reason to believe that the Arctic marine mammals were vulnerable given the amount of ice and frigid water that separated them from the rest of the marine world. With the exception of a handful of oceanographers such as Eddy Carmack, Humfrey Melling and Knut Aagaard, most scientists still regarded the Arctic Ocean as a marine environment unto itself with little connection to the north Atlantic or Pacific. What concern there was about disease was focused primarily on trichinella, a parasite commonly seen in polar bears but found to be increasing in walrus populations as well. Inuit hunters who eat either polar bear or walrus meat can get infected, sometimes with fatal consequences.

During those efforts to monitor the spread of trichinella in the 1980s, Ole Nielsen, a microbiologist who had been working on the identification and isolation of viruses in fish, transferred into the Department of Fisheries and Oceans Arctic Program to see whether anything else might be of concern to larger mammals. Nielsen was provided with hunter-killed and stranded marine mammal samples. His tests on those samples revealed the presence of brucella in walrus, beluga and narwhal all across the Arctic, including Foxe Basin and Repulse Bay.

Brucellosis is an extremely widespread disease that is of concern to both wildlife and human health. It has been linked to reproductive failure in dolphins and baleen whales, and there have been a few cases in which humans were made ill after coming into contact with meat from infected animals. While the percentages of infected animals were not high, they were sufficient enough to raise the eyebrows of people in the department and the microbiological community. One strain of the bacterium that Nielsen found had never been seen before.

Given the widespread nature of the disease throughout the world, Nielsen wasn't completely surprised to find it here in the Arctic marine environment. What troubled him though was how quickly the bacterium was spreading. He now sees it in 21 percent of the whales he tests, a fourfold increase from those early years.

Nielsen was surprised, however, to discover that neither beluga whales nor narwhal have antibodies that would help them resist distemper should the disease make its way to the Arctic. Like the Inuit and First Nations people who had never been exposed to the diseases that European explorers and settlers brought with them, these animals were potentially vulnerable. What this all meant for the future, no one knew for certain, but when Nielsen related these and other findings to a scientific conference in 2006 that brought Inuit hunters and scientists together in Tuktoyaktuk in the western Arctic, he was pretty blunt about the prospects for an ocean that was heating up so rapidly. "There are a lot of unknowns," he told me, repeating what he had told others at that gathering. "If distemper gets a foothold in the North American Arctic, it could get ugly. With there being as many as eighty thousand narwhal and one hundred and fifty thousand beluga whales living in the Arctic for most if not all of the year, a massive die-off somewhere is not out of the question."

A die-off in the Arctic is not as outlandish as it might seem. All that's required is a carrier: a pilot whale, harbor seal or dolphin—marine mammals that are known to carry the virus for long periods of time before suffering from the symptoms. Any number of them could ride a warm current of water into an Arctic marine environment that is no longer choked with ice. Killer whales are already doing it, so are harbor seals.

There's also the possibility that narwhal and beluga could come into contact with diseased animals on their annual migration south. Satellite transmitters attached to the narwhal Orr captured at Lyon Inlet the year before showed them swimming southeast through Hudson Strait toward the northern edge of the Labrador Sea, where distemper has already been isolated.

Disease is one of the tools nature uses to strengthen the resilience of birds and animals and keep spiraling populations in check. Every species in the Arctic carries a pathogen that has the potential to seriously undermine its health. Barren ground caribou carry brucellosis, mountain sheep and muskoxen have lungworm, polar bears and walrus are carriers of trichinella. Most of them get by in spite of it.

Although most diseases alone may not kill or seriously undermine the health of an animal, the dangers to health are likely to increase dramatically when chemical contaminants, industrial activity, nutrient level changes and new predators such as killer whales add to the stress that a narwhal, beluga or walrus already faces in the Arctic. Fisheries and Oceans scientists are beginning to see this kind of scenario unfold in the western Arctic, where mercury is washing out of the melting permafrost and showing up in the tissue of beluga whales.

So far, there is no indication that the beluga are being affected in a serious way by this mercury. But the Inuit of the western Arctic were concerned that was not the case for ringed seals, which, they complained, were in poor physical condition and no longer producing pups as often as they once they did. Lois Harwood, a Fisheries and Oceans scientist who has been working with Inuit hunters to solve the mystery, speculates that deteriorating ice conditions during the breeding season may be accounting for the drop in reproduction since 2003. As a precaution, she and John Alikamik, the Inuit hunter who has been assisting her in the field, sent samples to Nielsen to analyze.

Although not a toxicologist, Nielsen found no evidence to suggest that mercury or any other contaminant would account for this drop in reproduction. What he did find was a virus in the lymph nodes and lungs of some of the animals that he had never seen before. It took months of detective work and the help of several laboratories in the United States and Scotland to identify the three strains of picornovirus.

Picornoviruses can replicate in many species of mammals and birds. Highly diverse picornovirus-like viral sequences have been identified in marine seawater, as well as in a virus implicated in toxic-bloom-forming alga. But this picornovirus, which was the first ever found in a marine mammal, turned out to be distinct from any of them. When Nielsen and his associates submitted their findings, the American Society of Microbiology took notice and reported them in its scientific journal.

When I asked Nielsen if the seal picornovirus was capable of causing disease in humans and animals, he took some time to consider the question before answering.

"Is seal picornovirus capable of causing disease in humans?" he said, repeating my question. "I don't know. Is it responsible for the decline of seals? I don't know.

"All I know is that things can change and they can change very rapidly when it comes to emerging infectious diseases. Right now, I'm really the only person working in a marine area that extends from Alaska to Greenland. I need money and I need people if I'm going to keep on top of this."

Nielsen isn't likely to get money anytime soon. When the Canadian government recently earmarked $59 million for a National Aquatic Animal Health Program, he didn't get any of it. Virtually the entire amount was designated for fish farms and other aqua-cultural commercial endeavors on the Atlantic and Pacific coasts, which in themselves have the potential to spread disease. Still, he persists in doing what he can with what he has available and rarely gets questioned, even when he's called upon to sample specimens from places as far off his beaten path as California or Africa. "When you're walking down the halls with a bag of lion's guts, seal meal or whale blubber, people try not to engage you in too much conversation," he noted slyly.

Outside of what Nielsen is doing, what little disease research there is in the Arctic these days is being done by people such as Stephen Raverty of B.C.'s Animal Health Centre and Sandie Black of the Calgary Zoo, who was on this trip to Repulse Bay. Neither one's contribution amounts to a lot of time. Each year, Raverty donates a few weeks of his holidays to take up a cause he thinks is important to the Inuit. Black goes up for two weeks thanks to the goodwill of the Calgary Zoo, which is interested in expanding its Arctic program.

I found Sandie Black in the canvas cook tent at 3 a.m. writing a letter to her children. Slight but extremely fit, she struck me as the kind of person one is either likely to love or hate depending on one's ability to deal with people blessed with endless energy and infinitely sunny dispositions. During the first few days in camp, Black was tireless in her efforts be helpful in every situation that arose, be it unloading the boats, setting up the canvas wall tent or peeling potatoes and carrots in preparation for the evening suppers. Fortunately, if anyone in camp found Black's inexhaustible enthusiasm off-putting, he didn't show it.

As much as Black had been looking forward to this trip, she almost didn't come. Being a single mother and having to leave a young boy and girl at home with her parents was tough. So was leaving Maharani, the zoo's elephant, in the hands of her colleagues. Three years ago, the elephant had rejected her newborn and refused to feed it. In spite of zookeepers' round-the clock efforts to keep the baby alive, they were unsuccessful. Now Maharani was pregnant again and should have given birth by this time. Another death in the zoo family was the last thing Black or the zoo needed. It had not been a great year for Canada's second-largest zoo. Four gorillas had died in the previous year as a result of unrelated illnesses. The media scrutiny was relentless. Another death would raise more questions that could not be easily answered in twenty-second sound bites, even if there was a logical explanation for it, as there had been with the gorillas. Not surprisingly, Black had been on the satellite phone at least twice a day trying to keep up with what was happening down south.

"Crossing my fingers, that's all I can do from here," she said. "My boss urged me to go, and I know he's right, because we have good people at the zoo who are just as capable as me to deal with this situation. She really is in good hands. But still part of me wants to be there. It's hard to let go when you wait for more than twenty-one months to see this happen. Twenty-one or twenty-two months is how long a pregnant elephant goes before giving birth."

Black's boss, another veterinarian, got her involved in the narwhal project. He had worked with a Department of Fisheries and Oceans scientist to help tranquilize walrus during tagging efforts. Around that time, the community of Arctic Bay wanted to know what negative effects, if any, there were as a result of the capture and handling of their narwhal. When the department came looking for someone to do the physical assessments, Black's boss recommended that she go.

Black's story is typical of those who end up working with animals. She spent a good part of her spare time as a child watching chicks hatch from eggs in birds' nests and putting tadpoles into buckets so she could see them grow into frogs. She adored cats and dogs and was determined very early in on life to become a biologist. A friend suggested that perhaps veterinary medicine might be her thing.

"That's when it clicked that I wanted to be a vet," she said. "Being a vet was a way for me to marry my interests in the natural world with a need to be on the front lines of rehabilitation and the kind of research that's being done out here in Repulse Bay."

This was Black's fifth trip to the Arctic, so she had no concern about having to jump into the icy waters to help untangle a narwhal in the nets or possibly face a polar bear in the middle of the night while going to the one outhouse we had in camp. "It comes with the territory and the kind of challenges I face at the zoo," she said. "In my line of work, you have to be prepared for anything. One time we had this male lion that collapsed in the middle of its outdoor enclosure. My instinct was to rush to its aid. But that's not what you do with a lion that is half awake and in distress. So I loaded up a hypodermic dart to put him to sleep. Trouble is, it's hard to figure out how much drug will do the trick. It's also difficult to know if the drug has taken effect when you're as far back as we were. So we lobbed some stones at the lion to see if it was fast asleep. Then I walked in with two guys holding shotguns."

That said, most of Black's day at the zoo are filled with more routine jobs such as going over diets, monitoring quarantined animals and giving antibiotics to those animals that cut themselves or come down with some relatively minor illness. While going over the routine, she stopped to qualify this new line of thought. "I sometimes find myself taking it for granted until I realize that 'Hey, this is a grizzly bear that I'm working on here.' Or 'I'm out here beyond the north end of Hudson Bay catching narwhal.' Who can say that about their jobs?"

Black has little doubt that the narwhal in this region are doing fine, just as the Inuit from Repulse Bay think. But like Nielsen, she believes it's better to know what's going on now rather than later when there may be hundreds, or possibly thousands, of carcasses washing up on the shoreline.

"What do you do then?" she asked. "How do you deal with something like that in such a remote place? This, at the very least, will give us a baseline to work with. It will give us the heads-up in the event that warmer-water animals start moving north. There's a concern that they could introduce diseases that these animals have little or no immunity to."

Our camp was a luxurious one by some Arctic standards, even though it was situated on a pile of rock and sand that was virtually devoid of any kind of vegetation. The atmosphere was a lot more collegial than some camps I've been to where huge egos, impossible agendas and conflicting lines of responsibility have resulted in seriously frayed nerves, some vicious gossip and in rare cases a few punches thrown. Over the years, more than one person has had to be flown out of a research camp because of the stress that comes with the isolation, the constant cold, the polar bears, the mosquitoes and blackflies and dealing with type-A personalities that are overloaded with testosterone.

Each one of us had our own small nylon tent to sleep in, which was fine although not always restful when the loose ends of the flies were flapping in the brisk winds that were blowing in relentlessly. Unlike the glacier camp I had visited earlier in the year, we had a warmish canvas wall tent to go to when we woke up in the morning. We also had Joe Grande, the owner of one of Winnipeg's finer Italian restaurants, to work some magic on the two-burner Coleman stove.

Grande was as big as his booming Italian voice. He had come along as a volunteer to fulfill a childhood dream to see the Arctic. He was relishing the possibility of cooking up arctic char and other local food. But just in case, he brought along several bottles of homemade wine and a couple of coolers filled with Italians sausages, canned tomatoes, olive oil, garlic, spices, pasta and everything else he needed to cook the kind of meals he is more used to back home. He promised that we wouldn't go hungry and we didn't.

Given the very few opportunities we had to burn it off, it was a bit of a curse having so much good food. Most Arctic research camps have a strict rule about people venturing off on their own because there's always the danger of running into a polar bear. Ours was no different. We had the added worry of being away in the event that a group of narwhal got trapped in the net. Since narwhal need to breathe, getting to them as quickly as possible is imperative.

Once the camp was set up, most of the time was spent doing mundane things such as peeling carrots and potatoes, fetching freshwater from a meltwater stream nearby or playing bocce with the Inuit hunters who took to the ancient Roman game. When it was too cold or rainy to be outside, we slept, ate, read or drank copious amounts of tea.

As part of the daily routine, we were assigned a partner to work a two-hour shift every eight hours. The aim was simple enough; to be on the lookout for polar bears and any narwhal that might be heading our way. The lookout was especially important at nighttime because the twenty-four-hour Arctic days were now fading fast with each night that passed. The last thing Orr wanted was to wake up in the morning and find one or more narwhal drowned in the net because no one was there to see them get caught.

I was paired up with Jonah Siusangnark, a young Inuk from Repulse Bay. In his twenties, Siusangnark was no more than five feet tall and still living with his parents. Luky Putulik, one of the senior members of the Inuit party, advised me at the outset not to mistake his size for any lack of ability. "Jonah is a real hunter," he said, offering the highest praise an Inuk can give a fellow hunter. "A lot of people can go out there and come back with a caribou. But this kid could live out there if he wanted to. There are not many like him. Most people would be at home watching TV. Not Jonah."

It was tough going those first few days on duty. Most often we were alone at our post a few hundred feet away from base camp. On especially cold nights, it would be just me and Siusangnark standing outside with little to do but look to see if anything was moving in the water. Each time I tried to communicate, Siusangnark would smile, shrug his shoulders and go back to looking through his binoculars or spotting scope. It was shyness and a poor command of the English language that made him reticent, I think. But it didn't get any easier when Paul Tegumiar, his best friend, joined us as he often did when he was off his own shift. Tegumiar, I discovered, had barely uttered a word since his father passed away that winter and he wasn't talking on this trip. Whether this self-imposed silence was a show of respect or remorse, no one seemed to know or wanted to say. None of the Inuit men in our camp thought it unusual. They just kept on talking to him in Inuktitut and he'd nod or shake his head in response.

During those first few days, Jonah Siusangnark and I would sit on a tiny rise of land at our observation post and watch with great anticipation as one narwhal group after another would come straight toward the nets. On those rare watches when there was no wind or waves, we could hear the whales snort loudly whenever they came up for air. Each time, the narwhal would get to within a few feet of the net, they'd either swim around it or dive down and show up on the other side three or four minutes later.

It was uncanny how they did this but not surprising given the complexity of their communication abilities. Like beluga and other marine mammals, narwhal rely on acoustic signaling to guide them underwater to schools of arctic cod, away from killer whales and thick ice. They also whistle to talk to one another. Studies have shown that some pods have their own dialects. More recently, scientists from the Woods Hole Oceanographic Institution found a way of eavesdropping on three narwhal near Admiralty Inlet and discovered that each animal emitted its own distinctive whistle or pulsing sound. Ari Shapiro, the scientist who used suction cups to attach the recording devices to the narwhal, suggests that the whistling had more to do with communicating with other whales than foraging.

With a tusk that can be 10 feet (3 m) long or more, the male narwhal is one of the more extraordinary creatures of the Arctic. Luky Putulik told me how *Tuugaalik*, the Inuit word for narwhal, was created. According to legend, a long-haired woman refused to let go of a harpoon she had pierced into the side of a beluga whale. When the two got tangled up underwater, she continued to get wrapped up with the animal as it was trying to spin away. That, he told me, was how the beluga was transformed into a narwhal.

The outside world's take on the narwhal has been just as fantastical. In medieval times, many of the tusks that Viking hunting parties brought back to trade ended up in the southern Europe and the Far East. Invariably, those who were not aware of the origin of these spiraled tusks speculated that they were the horns of the mythical unicorn. This belief was shared by Martin Frobisher when he set off for the Arctic in 1576 in search of a Northwest Passage to the Orient. Frobisher also held a prevailing view that for every species of animal on land, there was a twin that lived in the sea. The narwhal was the sea's unicorn. Frobisher returned from his first trip to the Arctic with a tusk that he gave to Elizabeth I, Queen of England, in the sixteenth century. It was reportedly worth £10,000, the value of a castle.

Trade in narwhal tusks was profitable business for the Vikings, who brought many specimens, both single-tusk and the rare double-tusk variety, back from their forays into Icelandic and Greenlandic waters. The tusk had such universal appeal that it was used as a royal scepter in England.

The name *narwhal* is apparently a derivation of an Old Norse word meaning corpse. That may have referred to the animal's mottled gray coloring and to the fact that the narwhal sometimes lies motionless upside down in the water, resembling a drowned sailor after a shipwreck. It may also have something to do with Iceland's Pool of Corpses, where narwhal tusks were found at the scene of a shipwreck with the dead men floating in the water. Stories suggest that the narwhal were responsible for the incident.

Georges Louis Leclerc, Comte de Buffon and director of the Jardin du Roi in Paris, was undoubtedly aware of this story when he described the narwhal in his thirty-six-volume *Histoire naturelle*. The narwhal, he wrote, is an animal that "revels in carnage, attacks without provocation and kills without need or purpose." Linnaeus, the father of taxonomy, was the first to look at the narwhal through a more dispassionate scientific eye.

In *Systema Naturae*, his famous system for naming, ranking and classifying organisms, he listed the narwhal simply as *Monodon monocero*.

The narwhal goes by many names, including moon whale and polar whale, which was how early whalers referred to them. Its charm and notoriety has been enduring. In Jules Verne's *Twenty Thousand Leagues Under the Sea*, the submarine *Nautilus* is mistaken as a new class of narwhal. More recently, Chilean writer Pablo Neruda purchased a narwhal tusk with some of his Nobel Prize money and devoted parts of two poems to its name:

> You question me about the wicked tusk of the narwhal,
> and I reply by describing
> how the sea unicorn with the harpoon in it dies.

In another poem, he explains how his purchase bestows on him "the eternal dream of man!—health, youth and virility."

In spite of all this interest and infatuation with the animal, the narwhal remains somewhat of a mystery. Unlike the beluga, which thrives in captivity, every narwhal that has been placed in a zoo has died. No one has yet been able to explain why, nor can scientists explain why the narwhal almost never crosses a north–south line running approximately down the centre of the North American continent.

Scientists are almost certain that the narwhal tusk is not used as a weapon because specimens rarely show any nicks or breaks that might have occurred while fighting. Yet in October 1991, two Inuvialuit hunters from Tuktoyaktuk in the western Arctic found part of a narwhal tusk embedded in a beluga whale.

What we do know about the narwhal is that it is found mainly in the waters of Hudson Bay, the eastern Arctic oceans and the Arctic waters of Greenland. Growing up to 20 feet (6 m), it is slightly smaller than the white beluga, the whale it most resembles. Like the beluga, it dives to deep depths, up to 4,900 feet (1,500 m) or more in the narwhal's case. It can stay underwater for at least fifteen minutes. Narwhal migrate into Arctic waters in early summer when they follow the melting ice packs to feed on the arctic cod that abound along these edges. When the ice reforms in late fall, they retreat once again to the North Atlantic. Occasionally, they get trapped in small pools of water known as savsaats that become surrounded by thick ice. There, they become easy prey for polar bears and Inuit hunters who might happen to come upon them.

Males are born with an incisor tooth that can grow to lengths of up to 8 feet (2.5 m) in rare cases. Even more rare is a narwhal that has two tusks. These are the animals that are most prized by the Inuit because in addition to the substantial amount of food a narwhal offers, a long single tusk can bring in up to $3,000.

Martin Nweeia seems like an unlikely person to add anything new to the natural history of the narwhal. The dentist leads a pretty quiet life in scenic Sharon, Connecticut, where the town council counts the Annual Sharon Classic Road Race, the Christmas Tree Lighting and the Carol Sing as highlights in the lives of its three thousand well-heeled residents. Most of his days are spent filling cavities, making crowns and laboring over shade analysis for porcelain teeth. Once a week, he's off to Cambridge, Massachusetts, where he's a clinical instructor at Harvard University's School of Dental Medicine. His book, *The Whole Tooth*, is a no-nonsense tome that answers all the questions you might have about dentistry.

Idyllic as his life is, Nweeia hangs up his white lab coat and trades it in for a parka, toque and dry suit each spring when he heads to Greenland or Canada's eastern Arctic to focus on nature's biggest tooth, the tusk that protrudes up to 8 feet (2.5 m) or more from the jaw of a narwhal.

"It is, without a doubt, the most extraordinary tooth in nature," he told me in the early days of our stay in camp.

"Imagine a tooth that is half the size of your body. Not only is this the only straight tusk that we know of, it is also the only spiral tooth found in man or animal. It's a wonderful example of sexual dimorphism in teeth. You sometimes see it in females, but it's the male that has the tusks—the big one that you sometimes see sticking high up out of the water and a much smaller one that rarely grows to more than 30 centimeters [1 ft.]. In rare cases, you'll find a two-tusker, which is really remarkable to see."

We were sitting on that tiny rise of land that Jonah Siusangnark and I shared watching dozens of narwhal go past the nets. The tide was rising up behind us and rapidly closing off the only escape route we had to get us back to our camp on higher ground. But Nweeia was too animated to be concerned about the precarious nature of our situation.

"A few years back, I was up along the floe edge off the north coast of Baffin Island. It was early in the morning and the water was like glass. I had two hydrophones in the water so I could listen to the narwhal communicating. It was incredible. It was like I had walked into a large café. There were hundreds of conversations going on at the same time. 'Hi, how are you. Haven't seen you for a long time. How's your brother?' That's what I imagined they were saying. No. No!" he corrected himself. "I'm convinced that's what they were saying."

Nweeia had come along on this expedition in the hopes of solving a puzzle that has baffled scientists for more than a century. What value is there in a tooth of this size in a frigid world that is just a few degrees away from being covered in ice?

"The one theory that you often hear is that the tusk is used to joust with other competing males. Another is that it helps the narwhal break a path through the ice. But of all of the tusks I've seen over the years, and I've seen many, there's very little sign of scarring or chipping as you might expect from aggressive behavior.

"A third theory is that it is, like the lion's mane or the peacock's feather, a display tool used to get the attention of a female. The bigger the tusk, the theory goes, the more likely a female will be willing to mate." Nweeia doesn't discount this possibility. He does have a hard time accepting the fact that nature has gone so far out of its way to encumber just one animal in all of nature with a tooth used only for show.

Nweeia admits that the first few times he went to the Arctic, he wasn't sure where the research was going to take him. He talked to Inuit hunters, scuba-dived in icy waters, listened to the narwhal communicating and went home with some samples of narwhal teeth. Eventually, he called upon friends and colleagues to help him, including Fred Eichmiller, a dentist and engineer who specializes in developing materials for dental purposes.

Eichmiller wasn't even sure what a narwhal looked like when he agreed to have a look at what Nweeia brought back with him. What he saw under the microscope astounded him. Unlike human teeth, a narwhal's tusk is stubby at the root and tapers to a point at the tip. Instead of hard enamel on the outside and soft tissue on the inside, the narwhal tooth has a softer exterior that gives when the animal is jousting with another narwhal. At its centre, the narwhal's tooth is hollow like an old tree trunk.

In this long, hollow cavity, Eichmiller found nerves running from one end of the tooth to the other. The tiny tunnels that connected these nerves to the outer part of the tooth convinced both him and Nweeia that the tusk is used to detect sound, and sense barometer, temperature and salinity changes when sea ice melts or begins forming and when the weather turns for the worse.

"Think about it. Ice is both a threat and a refuge for narwhal," he told me. "Ice-edge environments are where arctic cod, the narwhal's principal food source, are found. It's also where killer whales won't go. But too much ice can also box narwhals in, leaving them vulnerable to polar bears or oxygen deprivation."

"This isn't as crazy it sounds," he said, sensing my disbelief. "Most teeth have what is known as a piezo effect. This is the voltage that is generated by a crystal when a mechanical force is applied to it. So when a twisted crystal such as the kind found in a narwhal's tooth moves with great force through the water and ice, it must generate some voltage."

In the hopes of testing this theory, Nweeia spent a year training how to use an electroencephalogram (EEG) so he could apply it to a captured narwhal while it was being tagged with a satellite transmitter. The EEG that he and Eichmiller brought along was no different from the ones hospitals use to monitor electrical currents within the brain. Electrodes are attached to the head of the whale. Wires attach these electrodes to a machine, which records the electrical impulses. The aim was to find how the narwhal's brain responds to different stimulants such as cold freshwater that comes from melting ice or warm salty water that flows in from the Atlantic.

"But first we have to catch a whale," he said, as we made a hasty retreat to get back to camp before the tides completely trapped us.

Most everyone was getting a little antsy by this, the fifth day of the trip. A year of hope and months of scrupulous planning for the live capture of ten whales was fading fast. So were the prospects for future funding. As difficult as it is to successfully conduct whale research in the Arctic, government and independent granting agencies aren't all that sympathetic to researchers who don't succeed in what they set out to do. Only Jack Orr, poker-faced and legendary for getting his whales at the last hour, remained calm—with a cigarette or two to help him.

In the meantime, we ate, slept and stood watch. With no place to go I was forced to get my exercise by pacing back and forth along a 650-foot (200 m) line that extended from one end of the camp to the other. Initially, this sent the two guard dogs we had in camp into a frenzy. After a while, though, they got so used to these marches that they slept right through the exercises.

Invaluable as these dogs should have been in warning us of impending danger, it was the sled dogs parked on the island across from our camp that announced the arrival of an intruder in the early hours of the following morning. At first, I thought it might have been an arctic wolf. The mournful call sounded very much like the long, low-pitched howl of an alpha male warding off an unwanted visitor. When I popped my head outside my tent, Jonah Siusangnark was standing there, looking through his binoculars. He turned to me and pointed to a spot at the end of the island, where a big polar bear had just emerged from the ocean and was shaking off the water that was streaming down from its coat. The sled dogs were going wild. We could see by the way that they were moving back and forth that they were more frightened than angered by the visit.

The bear just stood there for a minute or two sniffing the air before disappearing behind the island, leaving those of us who had been awakened to imagine what it was up to. Our thoughts were on the fate of the dogs; apart from the occasional yelp, they were mostly silent. Were they dead? Where was the bear? Would it swim over to us? After a half-hour of looking and listening, Siusangnark passed me his binoculars and pointed to the ocean. There, a half-mile (1 km) from the island, the bear was swimming out to sea. The dogs, in the meantime, resumed their pacing.

That night, a full moon emerged from behind the clearing skies and transformed the camp into a Halloween tableau. The white canvas cook tent, orange from the glow of lights inside, looked awfully forlorn set against a black sea. Siusangnark and I were on top of the hill coming to the end of a two-hour shift. Even with a -40°Fahrenheit (-40°C) parka, I was getting cold and more than ready to call it quits. Siusangnark was studying the moon through his spotting scope when he tapped me on the shoulder.

"Has anyone ever gone there?" he asked.

Siusangnark, I knew, from the increasing number of conversations we had, had never been to southern Canada. The farthest he'd ever been away was Rankin Inlet, a community of just three thousand Inuit—and that was too big and busy for his liking. So I assumed the question was sincere and told him the story of Neil Armstrong being the first man to visit the moon.

Siusangnark asked whether there was air to breathe. I said there wasn't.

He wondered whether it was cold. I told him that it was.

He speculated that maybe nothing lived there then. I told him that nothing did.

"Then why would anyone want to go there when they could come here?" he asked.

I didn't have a good answer.

Luky Putulik, Paul Tegumiar and Mark Tagornak were playing a game of bocce with Joe Grande the next morning when a big bull caribou was spotted on the hillside a half-mile (1 km) or so from our camp. Being low on freshwater, the trio used it as an excuse to go to the stream close to where the caribou was feeding. "Maybe we'll have caribou tonight," said Mark.

In the meantime, we sat and watched and waited for the narwhal, which we hadn't seen any sign of for more than twenty-four hours. I was beginning to wonder whether we would see any more. And then Laurent Kringayark pointed to huge numbers of gulls landing on the water in the distance. I saw them but didn't understand the significance. "They're catching cod, and where there are cod there are usually narwhal following," he said.

Fifteen minutes later, we could see dozens of whales moving our way. Everyone was feeling hopeful that this time we would get lucky. Once again, Kringayark signaled that something was coming, but this time from the other direction. He wasn't pointing to narwhal. It was nine boats filled with narwhal hunters, none of whom were apparently interested in what we were trying to do.

For the next hour, all we could do was sit back and hope that their gunfire might frighten one or more of the whales into the nets. I counted forty-two shots by the time it was all over. None of them were intended to hit the whales; the aim simply was to corral them close enough to the boats so that a harpoon could be thrust into the animals before they were shot and retrieved.

More than anywhere else in the Arctic, the hunters of Repulse Bay are encouraged to harvest whales in this manner so that an animal isn't lost. If a whale isn't shot in the spine or brain just as it is filling its lungs with air and a harpoon isn't there to retrieve it, the whale will sink. So many whales have been lost this way that one hunter told *National Geographic* photographer Paul Nicklin a few years ago that there was a fortune in ivory to be salvaged from the bottom of the Arctic Ocean.

The issue of lost and injured whales has become a delicate subject since the resumption of the bowhead whale hunt in the Canadian Arctic resumed in 1996. Repulse Bay was chosen as the site for the first hunt. The hunt seemed doomed almost from the start when some elders complained about the institutional nature of the operation. Many of them objected to the fact that the Canadian government was paying people to participate in the hunt while some locals were being excluded. Traditional hunts, they felt, didn't require payment for work nor the use of a special gun that sent big bullets exploding into the animal.

The day the first bowhead was spotted just outside of Repulse Bay, the whaling crew successfully harpooned the animal. Because no one had been properly trained to use the explosive device that was supposed to kill the animal quickly, the crew and bystanders—who were not supposed to be there—started firing with ordinary rifles. Badly injured, the whale swam out into the deep before it died and sank.

It was almost two days before the animal floated to the surface. By this time the whale was bloated with gas and unfit to eat. Except for a small amount of *muktuk*—whale skin and surface blubber—that was salvaged, the rest was allowed to rot on shore near the community.

Bowhead whale hunts since then have been more successful. The Inuit would like to increase the number they harvest now that it appears there are more whales than scientists previously thought. But in doing so, they are increasingly being challenged by animal rights groups and an urban world that no longer accepts the killing of any whales.

"You folks down south can kill your cows and pigs and eat all you want," Luky Putulik told me one night when I raised the issue. "But go to the store in Repulse and sec how much that kind of food costs. Most of us can't afford it. And it's not nearly as good for you as local foods are up here. Like it or not, this is our way of life. And nothing is going to stop us from doing it."

Shortly after sunset that night an unexpected opportunity presented itself. Laurent Kringayark, the captain of the Inuit capture crew, had been on the radio communicating with a group of hunters who had just harvested two narwhal. They were willing, he told those of us who were sitting in the cook tent, to have a couple of researchers come by and take some blood and tissue samples. That was all Sandie Black and Martin Nweeia needed to hear. They started packing their instruments for the long boat ride. Black was looking for samples to take back home and Nweeia was hoping for part of an embryo he could study to help him solve his puzzle.

Having sat for so long, I was desperate to go as well, even if my presence as a writer and photographer might not be welcomed. Jack Orr shrugged his shoulders when I made the request. "It's your call," he said. "If Laurent is all right with it, so am I."

Heading toward Frozen Strait, the channel of water that separates Melville Peninsula from Southampton Island, was quite the ride. The Arctic sun had already set behind the horizon and all that was left of daylight was a flush of luminous yellow and pumpkin-orange radiating across the western edge of a moonlit indigo sky.

Up front in the bow, Jonah Siusangnark, his back to us, used his arms to point and guide Kringayark around the shadowy chunks of ice that were being transported by the wind and tidal flows. Speeding across this Arctic seascape, it was difficult to stay warm and focused. When the silhouettes of several people standing on a small pan of ice 1 mile (2 km) from shore heaved into view, it felt like we were shifting back into a pristine world in which the relationship between humans and nature was much more simple.

Lamiki Malliki, the patriarch of one of the two Inuit family groups, greeted us. He nodded politely when I introduced myself as a journalist and then pointed to two live narwhals in the distance that were gently fencing with their long spiral tusks. Much relieved, I couldn't have expected a warmer reception.

Black and Nweeia wasted no time in dissecting what was left of the two whales that had been shot and butchered on the small pan of ice. The Inuit had already cut off the *muktuk* and were now piling the last of it into their boats.

"Nice shot," said Black as she pierced the thoracic cavity of one whale and watched a big pool of blood gurgle out around her. "This one died very quickly." Oblivious to the grotesque sight of the red-stained snow that was pooling around them, Black and Nweeia then started cutting and bottling the blood and tissue samples they were both after.

Standing there on top of the Arctic Circle, far from the industrial effluent that is killing beluga whales in the Gulf of St. Lawrence, I had a difficult time trying to imagine hundreds of diseased whales washing up in a place such as this, until I remembered Grant Gilchrist. He and his colleagues were on the other side of the island likely dealing with another outbreak of the avian cholera that no one expected would hit the Arctic.[1]

I thought about what Ole Nielsen told me the summer before. Things can change when it comes to infectious diseases, and they can change very rapidly—especially in a world that is warming up so quickly.

---

1. The disease struck the birds at East Bay five months earlier than the previous year. Far fewer died, partially because there were only half as many birds as the year before, likely due to 2006's avian cholera outbreak. Many of the eider that did return abandoned their nests when polar bears came ten days in a row to feed on the eggs. It was not a good year.

chapter six

WAKING THE DEAD

– Tuktoyaktuk, Northwest Territories –

The climate system is an angry beast and we are poking at it with sticks!

—Climate scientist Walter Broecker, 1998

THE WINDS WERE BLOWING IN FROM THE NORTHWEST, sending huge waves crashing into the small harbor at Tuktoyaktuk. A small group of us was standing on a hilltop along the shoreline watching the rollers wash over an artificial barrier of rock intended to stop the surge from taking another bite out of the hamlet and claiming it for the Arctic Ocean.

"In a few hours, that land you see out there will be part of the Beaufort Sea," said Jackie Jacobson, who had driven up in his pick-up. The Inuvialuit mayor was about to become a member of the Northwest Territories Legislative Assembly. "If it gets really bad, some of the roads will be flooded and a few people will be knee-deep in water walking to their houses. Unless we get some more rock to protect the point out there, that spit is going to be gone in a few years. Nothing's going to bring it back. That's when we really have to start worrying about the future. We need to get some rock in here to shore things up."

It was hard to believe that dozens of us had been on that spit of land the night before, racing canoes and eating arctic char, caribou and bannock. The feast had been cooked up by the organizers of the Coastal Zone conference that brought scientists, government decision-makers, Inuit and Inuvialuit together in a northern setting. The sea had been calm and nothing in the sky suggested a weather change that would bring in a storm of this magnitude.

Now, fewer than twelve hours later, dark gray clouds had blanketed all but a few of the fifty pingos that flank the outer boundaries of the hamlet. The giant cones of ice and soil, which can be as high as 161 feet (49 m), are more prominent here than anywhere else in the polar world. Although they look otherworldly, especially from the air, they form naturally when subterranean water percolates from the frozen floors of former lakes.

The wind was blowing so hard that we had to lean into it to walk. My plan to fly out to Herschel Island, an historic whaling station along the Yukon–Alaska border, had to be canceled. No one was walking the muddy streets of Tuk except to get from house to house.

"This is nothing compared to some of the storms that have hit us over the years," said Jacobson. "A few years ago, we had to evacuate people who were camped out at Shingle Point to the west, where many people from Tuk spend the summer. That storm flooded just about everything."

Situated at the north end of a vast stretch of treeless tundra that gradually slopes out from the Mackenzie Delta into the Beaufort Sea, Tuktoyaktuk is officially listed as being 16 feet (5 m) above sea level. That generous assessment does not take into account broad stretches of coastal shoreline, not much higher than sea level at high tide, which extend all the way into the Yukon and intermittently from there into Alaska.

Low-lying and sinking slowly into the sea, the western Arctic is very much like the lower Mississippi region around New Orleans on the Gulf Coast. The shoreline here has been slowly eroding for more than ten thousand years. However, sea levels are rising and the climate here is heating up faster than in any other part of the continent. Over the past century, mean annual temperatures have, on average, risen two to three degrees, with the greatest increases occurring during the past thirty years. That heat is now rapidly warming the permafrost that used to keep the ground glued together whenever the waves crash into the shore. Unlike New Orleans, where massive dikes protect the community from flooding, that pile of rock laid down some years ago is all that protects Tuktoyaktuk and its small harbor.

With sea ice and glaciers melting, polar bears starving, boreal forest fires burning out of control and warm water pouring into the Arctic Ocean, permafrost doesn't garner much attention from politicians, decision-makers or the media. But for half of Canada's landmass and most of Alaska and Siberia, it is the foundation on which towns, roads, pipelines and various ecosystems sit. As permafrost thaws, mountains and hillsides slide, swaths of forest fall down, houses and public buildings slump, pipelines crack and huge sections of roads turn into sinkholes. Inevitably, countless tonnes of carbon trapped in this frozen earth will be released into the atmosphere, further exacerbating the greenhouse effect that caused the thawing in the first place.

It's not just an Arctic phenomenon. This kind of thawing is also occurring at high elevations farther south where permafrost is imbedded in the earth. During the European heat wave of 2003, melting permafrost resulted in a massive rock fall on the legendary Matterhorn, forcing the evacuation of nearly one hundred people. The intense heat that summer also caused enormous chunks of ice to break away from a glacier above the resort of Grindelwald, Switzerland. Police were forced to stop tourists from entering the region until the wave of mud and rock that came tumbling down the mountain had settled.

In many places where permafrost occurs, there is no simple engineering solution to counteract the destructive impacts of the thawing. At last count, five communities in northern Canada and Alaska may have to be evacuated within the next fifty years if the big thaw in the Arctic continues to warm the permafrost and melt the Arctic sea ice. Another 175 communities are expected to suffer serious structural damage to roads, buildings and airports. Dozens of historic sites in Alaska and the Canadian North are also in danger of falling into the sea.

Where it's possible to address the problem, the price of doing something about it will be enormous. When the Canadian government considered the possibility of relocating Tuktoyaktuk several years ago, the cost was conservatively estimated at $50 million. Taking into account inflation and rising building costs, it would be at least double that today. In Newtok, Alaska, the U.S. Army Corps of Engineers recently pegged the cost of relocating the 315 residents living on the island at the mouth of the Ninglick River at $130 million, or $413,000 per person.

Back at the conference center, cleaning staff were busy using vacuums to suck millions of dormant flies that had been awakened by all the heat being pumped into the cluster of trailers that once housed oil field workers in the early 1980s. In one of the large meeting rooms that had somehow escaped the infestation, Steve Solomon and other scientists were doing their best to put the perils of climate change into perspective for Jackie Jacobson and the Inuvialuit who were in attendance.

Solomon is a soft-spoken man with dark eyes and a black, slightly graying beard that makes him look like an academic and somewhat out of place in a crowd of elders and hunters. Unlike many of the scientists and bureaucrats here who had no on-the-ground experience in the North, he knew the western Arctic landscape well enough to connect with the local audience.

For nearly a decade, Solomon had been monitoring the breakup of the Mackenzie River in springtime and then returning each summer to measure coastline erosion from the Northwest Territories to the Yukon. Big blows are not unusual in the western Arctic, he reminded everyone that day. Although those storms that cause serious flooding and erosion happen only about once every ten years, each one leaves its mark. The one that hit the community in 1944 flooded much of Tuktoyaktuk and swept away the old Hudson Bay Company warehouse. Another in 1970 killed two technicians who were on Tent Island in the Mackenzie Delta doing routine maintenance on a navigation tower. When the surge flooded the entire island, the technicians had no other option but to climb the tower and wait it out. Hypothermia did them in before the storm passed. The technicians probably never saw the storm coming. According to all the available data, winds were blowing 5 miles to 9 miles (8–15 km) from the southwest at the time they were on the island working. Five minutes later, they were gusting to speeds of 80 miles per hour (130 km/h), sending waves of water 3 miles (5 km) inland in some places along the Alaska, Yukon and Northwest Territories coasts.

A storm that Solomon witnessed firsthand in 2000 was mild by comparison. Yet it flooded the historic whaling settlement at Herschel Island, swept several important archeological sites into the sea and forced the evacuation of dozens of people camped at Shingle Point. By the time the storm of 2000 passed through, the ocean had swallowed up at least 23 feet (7 m) of shoreline around Tuk.

Being a city boy from New York state who had come to the wilds of Canada seeking fortune in the energy fields, not extreme weather, Solomon privately told me that it was not his place to tell the elders and hunters what they need to do. If, however, climate models prove to be correct, he pointed out, then future storms are likely to occur more often and pack a bigger punch than those that have hit the region in the past. The damage these storms cause will also be exacerbated by thawing permafrost, which is no longer holding the shoreline together as it once did. The erosion the storms cause will likely worsen as sea levels rise. The choices for the future, he said, will be slim. The government can relocate the community to higher ground or shore things up with a lot more rock. Either way, it will cost a lot of money.

To give people a better idea of what the future holds, Solomon produced a computer elevation model that digitally simulated what the powerful storm of September 1993 would do to the communities of Tuktoyaktuk, Aklavik and the seasonal hunting camp at Shingle Point if it were to occur again in 2050, when sea levels in the Beaufort Sea region are expected to be about 6 inches (15 cm) higher than they are now. Like the storm of 1970, wind speeds in 1993 peaked at close to 62 miles per hour (100 km/h). Solomon's colorful model neatly summarized how these strong winds, combined with rising sea levels and storm surges, would affect a low-lying coastline in the future.

Jackie Jacobson and many of the Inuvialuit hunters and elders in the room craned their necks to see what was on the screen. The 1993 storm surge water level, Solomon's model showed, was about seven times the normal tide range of 12 inches (30 cm) and about 8 inches to 12 inches (20–30 cm) higher than the 2000 storm. Not only would a future storm like this flood many parts of Tuktoyaktuk, it would sever access to the airport in five different places and put the community's freshwater supply in peril. The pile of rock we had seen earlier in the day would be completely overwhelmed.

Following Solomon's presentation, I sat down with a group of residents from Tuktoyaktuk who were weighing the pros and cons of staying put or relocating the community to higher, firmer ground. Chuck Gruben, an Inuvialuit hunting guide who was born and raised in Tuk, shrugged off the prospects of moving away. "We'll adapt," he said. "That's our way of life. Finding a way of living with the environment. It's always changing."

Fred Wolkie wasn't so sure. At seventy-two, the Inuvialuit elder had been around long enough to remember most of the big storms that had hit the Beaufort coast over the decades. He and his father, the son of a German/Swiss whaler who jumped ship in Alaska and made a life for himself in the western Arctic, had several close calls over the years. The storm that made the biggest impression was the one in 1944 that washed away several buildings.

"Charlie Rufus's boat went down with him and his family aboard in that storm," he recalled. "I was just a kid back then, and really only remember what we were told. But that same storm washed away most of the buildings on Baillie Island, where the Hudson's Bay Company had a post in the 1930s and 1940s. In the end, the Baillie Islanders all moved to Tuk because everyone realized that it was too dangerous to stay there. Many here are descendants of those people. They forget that."

Wolkie, who pronounces his name as "Wole-key," instead of "Vol-Kay," the way his grandfather did, wasn't alone in favoring the idea that the community be relocated. He suggested that they move inland to higher ground at Parsons Lake, an epicenter of energy industry activity in the western Arctic. "Better now when we can do it in an organized way rather than later when all hell breaks loose," he said. "And that's where all the jobs will be."

As the storm grew in intensity later that evening, Steve Solomon and his colleague Gavin Manson may have been tempted to hang around and join in the traditional meal that was being prepared for the scientist. There was, however, no other way for them to see what the waves and tidal surge were doing to the artificial barrier that we had looked at earlier in the day than to experience it firsthand. So they donned their toques, put on their raincoats and shrugged their shoulders when I asked if I could come along.

"This storm is nothing compared to the one in 2000, when 95-kilometer-per-hour [59 mph] winds were lashing the coast," said Solomon. "I had to sneak in between people's houses in order to be steady enough to take pictures of the damage. You could see by the way people were looking at me through their windows that they thought I was crazy.

"It was blowing so hard that this old boat, a derelict that was abandoned some time ago, got washed away in the surge that night. It was quite the sight watching it get spun around by the waves. That was the last time that boat went out to sea. All they found the next day were parts of an engine."

The storm of 2000, though not as powerful as the one in 1993, remains fresh in many people's minds, not only because it was so recent, but because it involved so much human drama. I knew because I was on the north coast of Banks Island that summer at the front end of the blow. Clouds of dust and sand turned the blinding sun into a rusty brown ball of fire for more than two days. The winds were so fierce that when the storm finally ended, my partners and I decided to knock down our tent camp and set it up just a few feet away rather than attempt to clear all the sand and dust that had penetrated everything inside.

After the storm passed, we flew back to Inuvik and bumped into Angus Simpson, a Parks Canada biologist who had just returned from a boat trip taking inventory of archeological sites along the coast of Ivvavik National Park in the northern Yukon. He vividly recalled the inky blue sky and pewter-colored waters that signaled something wild was coming his way. Spooky as it looked at the time, he and his colleague packed it in that night on the Yukon coast, thinking they would be able to ride out the big blow. When the first gust blew in and flattened their tent an hour later, they knew they were in trouble. Realizing how vulnerable they were on the low-lying coast with no harbor or higher ground behind them, they made a dash into the high seas to get to Pauline Cove, the protected harbor at Herschel Island.

Unbeknownst to them, Glen Gantz, a Utah biologist, was a little farther to the southwest, rafting solo down the Firth River on the Yukon–Alaska border. Having been delayed by logistical problems right from the start, Gantz was a day away from the coast and desperately trying to make up for lost time so that he would not miss the bush plane that was supposed to pick him up.

"I have never been so scared in my life," Gantz told me. "It took all the strength I had to get to the coast at the tail end of the storm when the rain turned to snow. What kept me going was the knowledge that there would be these old whaling shelters waiting for me there so that I could get warm."

When Gantz finally got to the coast, the full force of the storm had passed, but it was still snowing and blowing hard. He was cold, wet and in danger of succumbing to hypothermia. Much to his dismay, there was no sign of the shelters. Worried that he might not make it, he switched on his emergency locator transmitter in the hopes of getting rescued, not realizing that a plane or helicopter was unlikely to risk flying in for two hours from Inuvik to get him in that kind of weather. Fortunately for him, Simpson, who was by now settled in at the ranger station on Herschel Island, intercepted the distress call and decided to make the 3-mile (5 km) crossing to the mainland.

By all accounts, it was a daring rescue. Not until Gantz was plucked from his precarious perch did he learn that the night before he got to the coast, the waves had rolled over the spit and washed most of the windbreaks into the sea. Had he got there on schedule, as he had originally hoped, he would have very likely been swept away as well.

Although everyone on Herschel was safe, the island was hard hit. The Inuvialuit graveyard, one of four on the island, suffered the most damage. Some of the bones and clothes of people that had been buried over the past century were thrust up from the ground. The graveyards that entombed the bodies of whalers and RCMP officers fared better because of previous efforts to stabilize the area on which they sit.

The damage sparked a philosophical debate between young and old in the Inuvialuit community and between those who feel more connected to the past than the present. The Inuvialuit graveyard had been left untouched after the storm because of the elders' view that it is taboo to tamper with the dead. Better to return the spirits to nature rather than wake them up, they advised.

Andy Tardiff, the Inuvialuit park ranger at Herschel, was one of those who wondered whether it was prudent to leave the remains exposed. Tourists are now visiting Herschel Island by plane, boat and luxury ship. "From a spiritual point of view, I can understand this," he told me when I met up with him a few days after the conference ended. "But you can't just leave things exposed. People come here and walk away with things. Something has got be done about this sooner or later."

Herschel Island holds many other reminders of its rich and equally tragic past. The island's protected harbor at Pauline Cove is the only one between the Mackenzie Delta and Point Barrow in Alaska, which is why the island has been such an important stopping point for the Inuvialuit and for the nineteenth-century whalers who eventually settled here for more than a decade. During the high point of the whaling activity, as many as fifteen hundred people lived here. With little to do in the dark cold months of winter, many resorted to alcohol brought in by the shipload. More often than not, the flow of liquor led to rape, abduction, assault, murder and suicide. It is also resulted in the transmission of many diseases for which the Inuvialuit had no immunity. On top of that, the whalers literally exhausted the population of bowhead, beluga whales and caribou in the region by the time they cleared out for good.

Steeped in history, Herschel Island has been a leading contender to be Canada's next United Nations Educational, Scientific and Cultural

Organization (UNESCO) World Heritage Site. The last time UNESCO assessed the cultural importance of Herschel, however, the committee responsible warned that the historic value of the island may soon be "irreversibly compromised" if climate change continues to take its toll on the island. Whether anything can be done to save it in the long run is doubtful. Nicole Couture, a Ph.D. student who had been working on the island with McGill scientist Wayne Pollard, had spent several weeks on Herschel before stopping in at the conference in Tuktoyaktuk. The implications of this warming in the western Arctic was made clear to her and her colleagues when they watched in awe as huge chunks of permafrost—some 130 feet (40 m) long and 16 feet (5 m) wide—toppled off coastal bluffs before sinking into the sea. The storm that triggered this collapse was much milder than the one that was hitting the coast the last day of the conference.

Geographer Chris Burn was also on Herschel Island in the summer of 2007, continuing the ground temperature–monitoring program that he had started in 2001. He was reluctant to make too much of what his data was telling him given the short time since his study got started, but what he had collected suggested that the temperature of the permafrost had warmed by 1.5 degrees to 2 degrees Celsius. That was no surprise because he had already seen similar increases in ground temperatures in the Mackenzie Delta, where he inherited a monitoring program initiated by Ross Mackay, a University of British Columbia scientist, back in the 1970s.

Like Solomon, Burn is no stranger to fieldwork in the Arctic. In addition to the work he is doing on Herschel Island and the Mackenzie Delta, he has field sites near Mayo and Old Crow in the northern Yukon. His decision to get into permafrost science goes back to 1979, when he sat in on three lectures on the subject, one of which he distinctly recalls because it discussed the possibility that climate change might destabilize vast areas of frozen ground around the world. This lecture convinced Burn to do a master's degree and then a Ph.D. in this field of study. When the opportunity to work with Mackay presented itself in 1986, he never looked back. The two were still working together as recently as the summer of 2007, when they returned to the Mackenzie Delta that Mackay started working on in the 1970s.

Burn has no doubt that climate change is well under way in this part of the world. One can see it, he says, in the thick shrubbery overtaking the tussocky tundra in the delta and in the grasses now proliferating on the barren patches of ground on Herschel Island. Given the fact that permafrost descends to depths of 2,300 feet (700 m) or more in parts of the delta and other areas of the eastern Arctic, Burn predicts it will be thousands of years or longer before heat penetrates anywhere near those levels. That may not matter to engineers who will need to stabilize the top levels of ground that are now thawing. Sooner rather than later, says Burn, vast stretches of Arctic are going to become unglued.

Herschel Island is just one of hundreds of historic sites in the western Arctic that are being claimed or threatened by melting permafrost and rising sea levels. A few days after rescuing Glen Gantz, Angus Simpson stopped in at Philips Bay on the boat ride back to Inuvik. He had inventoried one of the archeological sites there on his way in to Herschel Island, and wanted to check up on it again. Two of the four sod houses that he had photographed just a few days earlier were gone and a half was missing on another. Thirteen feet (4 m) of shoreline had disappeared with them.

Over in Alaska, Benjamin Jones and his colleagues at the U.S. Geological Survey and the University of Cincinnati were seeing a similar situation unfold. Using aerial photographs that were taken of a 60-mile (100 km) stretch of coastline in 1955, 1979 and 2002, they were able to show how a dramatic increase in erosion in the last twenty-two years of that study period had destroyed a majority of the cultural and historic sites along that coast.

Farther east at Kitigaaryuit, where as many as a thousand Inuvialuit lived in what was once the richest and most populated area of Arctic Canada, the cultural loss will be even more profound. Kitigaaryuit is the name of an area at the mouth of the East Channel of the Mackenzie River. For more than five hundred years, the Mackenzie Inuit (Kitigaaryungmiut) gathered here each summer to hunt beluga whales. Those who stayed through the winter lived in elaborate multi-family, semi-subterranean sod houses that were framed with driftwood.

When a series of epidemics triggered by the arrival of the whalers killed hundreds of people at the turn of the century, the Kitigaaryungmiut moved away. Only a fraction of them returned when Anglican missionaries and then the Hudson's Bay Company moved into the area some years after. Long since abandoned, it is now a national historic site and a summer camp for beluga whale hunters. At last count, 190 graves, 17 sod-house ruins, the foundation of the Hudson's Bay Company store and other buildings on the site are in danger of eventually falling into the delta channel.

When Steve Solomon went in with cultural specialists Cathy Cockney and Elisa Hart several years ago to assess the situation, he and others concluded that nothing short of a monumental engineering or salvage effort could save the island's history from all the natural forces at play. Presented with that eventuality, the elders once again chose to allow nature to take its course. Like Andy Tardiff, the ranger at Herschel Island, Cockney, an Inuvialuit woman with deep roots in this part of the world, was torn by that decision. So was David Morrison, an Arctic archeologist with the Canadian Museum of Civilization, who has spent a good part of his career excavating sites such as Kitigaaryuit in the western Arctic. But Morrison was resigned to this fate. The entire coastline has been sinking into the ocean for so long, he told me, that it's probably too late to do anything now. There are few sites left that are more than fifty or one hundred years old, he notes.

Saving history isn't the primary reason why Solomon and his colleagues have been coming up to the western Arctic in recent years to get a better handle on storms, rising sea levels, melting permafrost and what's happening on the floor of the Beaufort Sea. Oil and gas exploration has really been driving the research.

The Mackenzie Delta and Beaufort Sea are quickly becoming the next major energy frontiers in North America, if not the world. Since the

1970s and 1980s, when more than ninety offshore wells were drilled, at least twenty-one significant discoveries have been made. Most everyone in the industry is so confident that more will be found that plans are well under way to build a $16-billion natural gas pipeline up the Mackenzie Valley.

But there are still many unknowns about how ice scouring, mud volcanoes, erosion, melting sea ice and future storms will affect these future developments. "Any company that has plans to build a pipeline, a gas processing plant or oil refinery will not want to play the odds on erosion," Solomon told me. "With parts of the Mackenzie Delta and Arctic coast eroding at rates as high as 10 meters [33 ft.] a year, the engineering questions become very challenging. Where and how deep do you bury an offshore pipeline? If you're going to build a processing plant, where do you put it so that it won't be vulnerable to storms and erosion?

"Understanding how storms have affected the region in the past is helpful. But you can't rely on the past to answer questions about what might happen in the future. There are so many variables to consider and not all of them are reliable. There is still a lot of work that needs to be done in this area to answer these questions."

Nightmare that it promises to be for engineers and frontier energy producers, the thawing-permafrost story could have disastrous implications for the entire planet. With so much organic matter trapped in the permafrost, some scientists are likening the current thawing to a "carbon bomb," which will eventually trigger the release of frozen greenhouse gases that have been freeze-dried in the soil for several thousand years. Cracks that form in the thawing permafrost could also provide pathways for huge volumes of methane gas trapped in the ocean bottom and in ice crystals in the permafrost to rise up and escape.

Methane is of particular interest to scientists and climate modelers because it is twenty times more effective than carbon dioxide in trapping heat in the atmosphere. A sudden rush of methane into the atmosphere would expedite the greenhouse warming that is already well under way. Some scientists conjecture that the last time this happened was 55 million years ago, when the Arctic was inhabited by the alligators, tortoises, lemurs and other warm-weather animals that Mary Dawson found on

Ellesmere Island in 1975. It added warmth to an Arctic that was beginning to cool off after the era of the dinosaurs ended. As unlikely as this sudden rush of methane is to occur again in the near future, a team of scientists on contract with the U.S. Geological Survey warned in 2008 that the emission of methane from warming permafrost, gas hydrates and wetlands will likely have a significant impact on global warming for the next one thousand to one hundred thousand years.

Energy industry experts are also intrigued because methane is mostly natural gas, which can be sold as energy. By all accounts, there is a tremendous amount of methane stored in gas hydrates, bands of highly concentrated methane gas frozen in the permafrost. Gas hydrates, latticelike ice structures that trap large quantities of methane, are found on both the Atlantic and Pacific coasts of North America and in many others parts of the world, but those located in the western Arctic are among the most accessible and potentially the most economical to exploit. Scott Dallimore, a Geological Survey of Canada scientist who is heading up an international effort to find a way of extracting them from the western Arctic in a commercially viable way, likens the energy potential to the tar sands in Alberta.

In some places in the Mackenzie Delta, shallower reservoirs of methane are naturally percolating to the surface. Whether this is caused by permafrost being thawed by rising temperatures and warmish seawater slowly moving inland, no one knows. Scientists Rob Bowen and Michelle Cote, on contract with the Geological Survey of Canada, have been putting together a profile of methane seeps in lakes and ponds on the Mackenzie Delta. What they've found so far is sobering.

Bowen thinks it's wrong to characterize these as mere seeps. Considering how much pressure is driving them to the surface, he likens them to hot-tub jets. Three of the biggest seeps he and Cote have measured produce as much greenhouse gases in a year as that emitted by nine thousand average-sized cars. If you flared them, he told me, they'd send flame 33 feet (10 m) into the air. They're so vigorous that in most cases they prevent ice from forming over them, even in the most extreme months of an Arctic winter.

Bowen found this out the hard way earlier that spring a few days before I met him in Inuvik. On site, he made the mistake of using an ice auger to try to drill into one seep that tends to freeze over. "After drilling in about a meter [3 ft.] I could feel and hear the ice rumbling beneath us," he told me. "Realizing that a large pocket of gas had pressurized under the ice plate, I yelled 'run.' Immediately after, a mixture of water and methane shot up into the air about 10 meters [33 ft.] like a geyser, soaking me completely. I was lucky that the helicopter was standing by so I could change and warm up."

Bowen and Cote are now using high-resolution satellite imagery to try to determine how many methane seeps there are in the delta. Although he has no idea what the number might be, he suspects it will be significant and likely to increase in the future. "A big part of the Mackenzie Delta is very close to sea level," he said. "If sea levels rise in the future, much of the delta will be flooded. If this is the case, it will be interesting to see if the warming effects of the ocean will result in further degradation and stability of the permafrost. Theoretically, that could result in more conduits that will allow methane gas to rise to the surface."

Should that happen, then forecasts for future warming will have seriously underestimated what's to come. None of the climate models factor in the methane gas that could be released in the Arctic world and from the ocean floor. They also do not take into account the greenhouse gases that are locked in the soil.

Standing on the coast that evening, I watched with some wonder as Solomon and Manson casually went about their business. With the winds moaning and the waves crashing onto shore, there was little point trying to communicate in words. Both moved about effortlessly, taking photographs and recording satellite transmissions as if they were in perfect synch. Like sailors on a ship in a storm, neither one seemed troubled by the wet weather or the cold temperatures.

Listening to the travails of other scientists at the conference, I got the clear impression that those studying the world of permafrost are a rare breed. Part of this resilience, no doubt, comes from the nature of the work in which permafrost scientists are engaged. When not drilling holes in intense cold, struggling to avoid frostbite, they are drilling holes in stifling heat, fighting off biting flies and mosquitoes.

Not that this doesn't send some of them dangerously close to the edge—but veterans have ways of coping. Some play games such as "Page Watch," where the goal is to kill as many mosquitoes and blackflies as possible by slamming a field book closed in the most timely and deadly fashion. The victor is the one with the most dead insects flattened on a page.

Permafrost scientists are such brutes for punishment that Peter Kershaw, a University of Alberta scientist who has devoted his entire career to the study of frozen ground, had the dubious honor of being named to *Popular Science* magazine's list of the "Ten Worst Jobs in Science" in 2005. He finished just ahead of the orangutan-urine collector in Borneo and two places behind the human-semen washer in Los Angeles. The same year, however, Earthwatch named him, above the 130 scientists the organization supports in the field, as Principal Investigator of the Year.

There is a touch of daring genius in this group of men and women who work in the Arctic. Ask anyone why and, invariably, they'll tell you it was because scientist Ross Mackay, Chris Burn's mentor, set the bar so high in 1978 when he came up with the big idea of draining an entire lake in the Mackenzie Delta to better understand permafrost processes.

The boldness of such a plan would probably never get by an environmental review today. Not only did that gambit open the door to a whole new field of research in permafrost science, it established Mackay and his colleagues as world leaders in the field. During his long career, Mackay single-handedly published one hundred and fifty papers on the subject and co-authored another fifty. Since then, permafrost scientists have aspired to emulate him and his accomplishments.

If Ross Mackay was the prince of permafrost science on land, Steve Blasco is, in some ways, his twin in the study of permafrost and other geological features on the ocean floor of the Arctic. I understood why when I sat in and listened to him at the conference as he somehow managed to turn a talk entitled "Contributing to the Sustainable Canadian Beaufort Sea Hydrocarbon Development through Seabed Geoenvironmental Research" into a rollicking adventure story chronicling his discovery of underwater mud volcanoes in the Beaufort Sea. Although Blasco's discovery is undoubtedly invaluable to oil and gas experts trying to minimize the risk of drilling in the Arctic Ocean, his lecture was also a fascinating geography lesson that would have left Discovery Channel producers beaming had they filmed it for television.

This wasn't the first time I had seen Blasco in action. We'd run into each other several years earlier on the *Nahidik*, a Coast Guard vessel that was helping him and other geologists map the floor of the Beaufort Sea. Blasco was just as entertaining back then as he was this time around, regaling me and others with stories of research trips that took him to the North Pole, the Caribbean, the Great Lakes, the Russian Arctic—where he had a personal meeting with First Lady Yeltsin—and deep down to the depths of the Atlantic Ocean where the *Titanic* now rests. Blasco was an adviser on the IMAX film *Titanica* and the first to test the unmanned submersible TROV (Telepresence Remotely Operated Vehicle),when it was used to study geological and geophysical properties of the seabed in the Arctic. In 1995, he was part of an expedition that used underwater sonar to search for the lost Franklin expedition.

What's most remarkable about Blasco is that he is legally blind but somehow able to fool most people into assuming that he isn't. During his presentation in Tuk, I watched as people listened intently, taking no notice of his big eyes darting back and forth behind thick glasses that capitalize on what little peripheral vision he still enjoys.

"No one knew what the ocean floor looked like, so I was no worse off than anyone else when I got around to doing this sort of thing," he told me when I asked him later how he got into this field of study. "It's a strange world down there with all that ice scouring and mud volcanoes on the sea floor venting gas. Some parts of the Beaufort Sea are almost sterile. Then you get these hot spots that the biologists are interested in. Surprisingly, some of them occur in those areas where former oil and gas ice islands have been bulldozed. It's important because if there is going to be drilling for oil and gas, you want to be able to protect some of these hot spots. You also want to know what areas are vulnerable to ice scouring."

With his white beard, white hair, big thick glasses and a frame to match Santa Claus, Blasco looks like he'd be comfortable spending so much time in the Arctic. "Much as I'd like to think that," he corrected me, "it's just not true. I counted the days, the minutes and the seconds on that trip to the North Pole—'62 days, 12 hours, 19 minutes and 35 seconds'—and I don't know that I was ever warm. I brought my wife up here this time around because she told me not to come back if I didn't take her. I like to think she's keeping me warm. She's always wanted to see the Arctic, but I'd just as soon be in the Caribbean."

The Arctic, he admits, beguiles him nonetheless, as it does so many others who have spent time there. Like Solomon and Burn, he believes that researchers like him really need to get a better handle on what Mother Nature is doing in the Arctic if engineers are going to find a way of protecting low-lying coastlines and other vulnerable areas. "Each year of warming, it seems, the reach of a storm extends farther inland," he said. "People down south might not realize it, but much of what is happening in the Arctic now is going to hit them in the future. It may not hit them as hard, but it will make them take notice."

A week after the conference ended, I was in Aklavik, where 750 people of Inuvialuit and Gwich'in origin live 70 miles (113 km) from the coast in the heart of the Mackenzie Delta. The only highway that connects the town to the outside world is a winter road that runs along the ice on Peel Channel for half the year. Like other Mackenzie Valley communities, people here are concerned that the ice road to the outside world will no longer be reliable for as long in a year as it once was.

Although it is far inland, Aklavik is more vulnerable to permafrost thawing than Tuktoyaktuk because of the ice jamming on the river that tends to cause the banks to flood over in spring. At one point in the 1950s, the Canadian government was so certain the town was going to sink into the delta that it tried to get everyone to move to Inuvik, the new town it built along the delta's east channel.

Danny Gordon was just ten years old in 1948 when his family moved to Aklavik after traveling by dogteam from their home on Barter Island in Alaska and settling for a while on Herschel Island. Life was too good to tempt his father to move and start all over again in Inuvik. So, like others who protested loudly when the government told them to move, they ended up staying.

After sixty years of navigating effortlessly through the huge spiderweb of channels that connects many of the forty-five thousand lakes in the region, Gordon figured he knew every inch of the delta. But when I met up with him that day, he admitted that the mental map he's relied on to get him from one body of water to another was becoming a puzzling maze that no longer leads him so readily to the grizzly bear, moose, muskrats, lynx, mink and other wildlife that he's after.

"Riverbanks are slumping, channels are changing and some lakes are disappearing," said Gordon. "You don't find animals where they used to be. The delta is not the same place it was twenty or even ten years ago. It's changing, big-time."

If Environment Canada scientist Philip Marsh is right, the changes Gordon is seeing are just the beginning of even bigger ones coming down the road. He and Lance Lesack of Simon Fraser University have been studying the delta for the past quarter-century. In addition to more thawing of the permafrost that is causing shorelines to slump, they predict that as many as one-third of the forty-five thousand lakes could, under certain conditions, eventually dry up.

Most of the lakes are very shallow. The one-third of these that are most vulnerable are situated at higher elevations. If spring floods fail to reach those higher elevation lakes, as Marsh and Lesack predict, most of those fifteen thousand lakes will disappear fairly rapidly, within just ten years.

The Mackenzie is no ordinary delta. Not only is it the largest intact delta in North America, it is twelfth largest in the world. It is so big that the 230-square-mile (600 sq. km) Kendall Island Bird Sanctuary, summer home to sixty thousand nesting shorebirds, represents less than 5 percent of the area it covers.

The fact that the delta world is changing should come as no surprise given what's happening on the coast. But Marsh now believes that he and his colleagues have underestimated just how fast and how dramatically it could all unfold in the future. Their forecast is based on a hypothetical scenario that is contingent on temperatures increasing, precipitation remaining relatively the same and there being less ice-jamming along the river.

Although the model is hypothetical, the predicted changes are already happening. Not only are flood-causing ice jams occurring earlier than normal in spring, but the volume of water that backs up when they do occur is often not enough to fill the channels that connect the high-elevation lakes to the river.

A loss of fifteen thousand lakes would not only be catastrophic for birds and wildlife, but also for Aboriginal peoples such as Gordon, who depend upon wildlife for a living. Both Inuvialuit and Gwich'in hunters have already accepted the fact that natural gas and pipeline developments in the delta are going to result in some undesirable changes in the environment. In the Kendall Island Bird Sanctuary area alone, there's a plan in place to build an airstrip, seventeen gas wells and a camp that could serve as many as three hundred people.

No one, however, has quite come to terms with the possibility that fifteen thousand lakes will disappear. Gordon wouldn't be surprised to see it happen. He just doesn't want to be around when it does. "We haven't seen -50°Celsius [58°F] for twenty years," he told me when we parted. "It used to be that we had five or six feet [1.5–1.8 m.] of ice covering the lakes in winter. Now, we don't get more than three feet [1 m.]. Summers are also a lot hotter. It's a different world up here, no doubt about it."

chapter seven

IQ

– Off the Coast of Northwestern Hudson Bay –

I went from the iglu to microwaves in less than forty years. I lived and saw the end of the traditional way of life and the beginning of the modern technological age. In the years of change, I have come to realize that the Inuit way of life is disappearing and that my Inuitness is becoming more Europeanized or Southernized. Our struggle is to protect values and ways of thinking that are different from southerners, but they have a lot to offer to a world suffering from environmental, social and political pain.

—Peter Irniq, Inuit cultural teacher and philosopher, 2006

Marble Island looked like the back of a big white whale heaving in and out of view as the warm spring air bent the last frosty rays of sunlight over western Hudson Bay. All day long, Gabriel Nirlungayuk resisted the temptation to lead me on our snowmobiles on the 9-mile (15 km) sea crossing over to this dazzling outcrop of white quartzite. Because of shorter winters, the solid ice that traditionally dominates this seascape is not the forceful match it once was for razor-sharp winds and powerful tides. The salt water no longer freezes over completely in many places. Thin ice and open leads may be good for whales, walrus and seals that need a place to breathe in the cold dark months of winter, but they can be dangerous for an Inuk like Nirlungayuk, who prefers a solid platform on which to travel.

Since the vast sheet of year-round ice began retreating from this part of the world approximately seventy-five hundred years ago, Hudson Bay has been fertile ground for Inuit hunters living along the shores of Nunavut and the northern shores of Manitoba, Ontario and Quebec. Seven months of snow and ice cover still keeps it cool long enough for polar bears, barren ground caribou, beluga whales and other Arctic species to thrive at latitudes that normally favor black bears, white-tailed deer, killer whales and other animals.

Hudson Bay, however, is warming up. In the past decade, capelin have overtaken arctic cod as the region's main fish. Brook trout are making their way up the western coast. Killer whales, once blocked by the ice that choked Hudson Strait, are beginning to prey on beluga and narwhal. Unable to hunt seals for as long as they were able to do in the past, polar bears are declining precipitously in number. Diseases such as avian cholera are beginning to take their toll on common eider duck colonies and, potentially, other Arctic birds and animals. Virtually every facet of Inuit life is being affected by this warming. Not only are some hunters finding it more dangerous to travel in springtime, but many are having difficulty tracking down the caribou and other animals because their migration routes are changing just as their numbers are dropping.

Much has been made by Inuit leaders, academics and government policy-makers about the changes taking place. Although most everyone agrees that something needs to be done to help preserve the Inuit way of life, they have struggled to come up with practical solutions. More often than not, science is taking a backseat to the kind of traditional knowledge that has guided the Inuit for thousands of years. The Inuit's "right to be cold," I was about to learn on this trip, is a vexing problem that won't be solved simply by putting polar bears, walrus and other Arctic animals on the endangered species list.

It had been just two days since Gabriel Nirlungayuk and I left the Inuit community of Rankin Inlet to hunt for walrus and seals. By this point, I was already beginning to wise up to the difficulties that the Inuk face each time they venture out into the cold with just a snowmobile, boat, rifle, stove and a tent to put meat and *muktuk* on the table.

Nirlungayuk and I had met the summer before at the Coastal Zone conference in Tuktoyaktuk. "Trust me," Nirlungayuk had said, challenging me to visit him at a time of year when the Arctic can be more sublime and repelling than picturesque and inviting. "After you spend a week on the ice in late March, you'll understand why it is that many of us are tired of people from down south telling us what to do."

Although it was early in April that I took Nirlungayuk up on his challenge, the weather was still acting as if it was early March. The warm breeze that had blown in was merely a brief puff of southern air that vanished with the next northwesterly gust.

It was late in the evening and there had been no sign of the carnivorous rogue walrus that had surprised and nearly taken the life of an Inuk elder we had met hunting along this floe edge earlier in the day. Although we had seen several big, bearded seals basking in the sunlight on day-old grease ice, it was too thin for us to travel on. All we could do is gaze at them through binoculars. Nirlungayuk had no desire to launch the aluminum boat that he had set up on skis behind his snowmobile and risk the chance of being crushed or capsized by moving ice.

Now that our long day was nearly over, I was looking forward to supper at camp—even if it was going to be more of the fermented narwhal *muktuk* and arctic char Popsicles we had gnawed on at lunch. When Nirlungayuk suddenly turned his snowmobile into the setting sun, I realized I would have to wait a little longer.

To rid himself of the temptation of crossing over to Marble Island on thin ice, Nirlungayuk was apparently giving in to it. So with me following closely behind on my snowmobile, we started making our way around the jumble of ice floating in the open water toward the island that had haunted us all day.

For a time it looked like we wouldn't make it. It was weird: the more we pushed toward the island, the farther it seemed to recede into the distance. Ice fog rose from the dark, lacquered surface of the open lead nearby, as though something far below the ocean depths was exhaling a warning, telling us to turn back. I was further unsettled by the noises emerging from the growlers—large pieces of ice that stick 3 feet (1 m) up from the water. Sometimes when I stopped to figure out what lay ahead, I could hear the trapped air inside the ice escape. When it didn't pop or crackle, it sounded like an animal growling.

Unnerved, I torqued the engine, hoping to narrow the distance between Nirlungayuk and me. But before I had time to catch up and signal my desire to retreat, the faint shoreline of Marble Island suddenly came into sharp view. When I finally joined Nirlungayuk on the island, he didn't welcome me the way I had expected.

"Get down on your hands and knees," he said, motioning for me to do so quickly. "You must get down."

Assuming there was a big polar bear somewhere on the hillside, I quickly did as he ordered, but hard as I looked, I could see nothing but snow and white rock around us.

"Some people believe this place is haunted," Nirlungayuk explained while I lay there on my belly. "It's customary to crawl the first time you visit this island. This ensures that you will not be afflicted with the sickness and death that this place has brought to so many in the past. The ghosts are watching us."

In some ways, the ghosts of Marble Island illustrate why the challenge of climate change has become such a formidable problem for Inuit leaders and government policy-makers. Like the scientists and Inuit hunters who see climate change in different ways, the spirits that dwell here come from two very different worlds.

One of the spirits belongs to an old woman who lived along this coast of Hudson Bay some time ago. When famine forced her son and his family to find food elsewhere, she refused to come along, lest she slow down their desperate search. Left alone, the woman used her shamanistic powers to turn a large block of sea ice into the island that it is today. When the family returned to the area years later, they found no physical trace of the matriarch. But while they were searching for her, they—and others who have come here since—could hear her voice in the wind, whispering, "Don't worry. I have got my wish. My spirit now lives on this white island."

The other spirits belong to the crew of two ships, one captained by James Knight, the other by David Vaughan, who took refuge here in 1719 on a Hudson's Bay Company search for gold and copper and a northwest passage to the Orient. The shearing force of incoming, tide-driven ice damaged the ships irreparably, leaving all forty crew members stranded. Only five made it through the second summer.

According to the Inuit who eventually arrived on the scene to find the last two survivors, the men were so sick and weak, they were unable to digest the seal and whale blubber they were given. In the throes of despair, the two huddled together and wept bitterly. When one finally died, the other did his best to give him a proper burial. Even this simple task proved to be too much. The last man breathing lay down and died shortly after.

We know all this from eighteenth-century explorer Samuel Hearne, who got the story firsthand from the Inuit nearly fifty years later. What remains a mystery is why none of Knight's men went across the ice to the mainland to barter for food, clothing and whatever else they would have needed to walk back to Fort Prince of Wales, from where they had set sail.

Like many others who know the story, Nirlungayuk thinks he understands why the Englishmen were reluctant to reach out to the Inuit. "They didn't trust us," he said as we walked toward a wooden cross marking the grave of one of Knight's men. "They had dealt with the Dene, our bitter enemies back then, for so long that they refused to make contact. They thought we were ignorant barbarians, that we knew nothing that was of importance to them."

Nirlungayuk believes that southern-based scientists and politicians involved in Arctic policy decision-making are sometimes guilty of the same inaction. Like many Inuit hunters, he believes that the traditional ecological knowledge the Inuit have learned over thousands of years of living in the Arctic is either being ignored or undervalued when it comes to dealing with climate change issues.

As the director of wildlife for Nunavut Tunngavik Inc., the Inuit organization that oversees the settlement of the 1993 land claim that gave the Inuit of the eastern Arctic legal title to 135,522 square miles (351,000 sq. km) of land and mineral rights to 14,286 square miles (37,000 sq. km) of that territory, Nirlungayuk mediates between Inuit hunters who believe in traditional knowledge and university-trained scientists who sometimes see the natural world in a very different way.

Because he is a believer in both science and traditional knowledge, Nirlungayuk is not in an enviable position. Increasingly, Inuit hunters in the Hudson Bay region have been up in arms about scientific reports that say polar bears are declining in at least three regions of the Arctic, including southern and western Hudson Bay. Some hunters were threatening to ignore the limits on polar bear hunting and take as many animals as they pleased.

The contentious issue of tranquilizing and collaring animals for research purposes is also heating up. Many Inuit hunters believe the drugs are tainting the meat and affecting animals negatively. Some are so upset with scientists that the hunters have been talking about putting an end to the practice.

The Inuit's fears are not entirely unfounded. Although all evidence so far suggests that the drugs are excreted within weeks of darting an animal, Health Canada still recommends that the Inuit refrain from consuming any creature that has been tranquilized in the past year. One study also demonstrates that tranquilizers take their toll on grizzly bears down south, but similar effects on polar bears have not been documented.

In threatening to put a stop to the scientific research, the Inuit are risking sanctions by the United States, which has the power to restrict the movement of animal parts from the Arctic regions into its territory. The

Americans at the time were already well on their way, listing the polar bear as a threatened animal. Doing so would not only strike a blow to the lucrative sports hunt in Nunavut and the Northwest Territories, which brings $3 to $4 million of revenue to Inuit communities annually[1], but it would undermine a way of life already in trouble for other reasons. For years, alcohol, drugs, satellite television and resource development have been taking their toll on language, traditional family units and the survival skills that define an Inuk. More often than not, young people are turning their backs on the traditional ways that made the Inuit so comfortable living in such a harsh environment.

"You're not going to find many of us saying that the Arctic isn't warming," he said after we walked to the top of the hill and turned our sights toward a huge body of open water on the other side of the island. "Ask anyone and they'll tell you that animal migrations are changing, that it isn't nearly as cold as it used to get and that the ice is getting thinner and melting faster in spring. You only have to look out there to see that.

"But the fact is the Inuit have adapted to climate change like this in the past, and we will continue to do so in the future. We know better than anyone else what's happening here. We should be the ones making decisions about the future."

No one who has traveled with experienced Inuit hunters in the Arctic can deny they have a deep understanding of the world they live in. The best are guided by water skies and ice blinks that might otherwise be mistaken for stormy weather or clear skies approaching. They know that ringed seals build subnivean lairs in snowdrifts along the lee and windward slopes of pressure ridges and that beluga whales come in by the thousands in July and August to exploit warm freshwater estuaries such as those found at the mouth of the Mackenzie, Cunningham and Nelson river systems.

---

1. Shortly after this trip, the U.S. government did list the polar bears as "threatened." The impact on the sports industry has not yet been measured.

Invariably, an Inuk like Nirlungayuk will see an animal with his naked eye long before someone like me from the south spies it through binoculars. I was reminded of this on the hilltop that evening when Nirlungayuk turned toward the mainland and pointed to some black dots in the distance. "Maybe we'll have caribou for supper tonight," he said as I did my best to focus on the herd of animals that I could barely recognize without the spotting scope. I estimated that there were at least two hundred of them digging into the hard snow to feed on the sedges and lichen below.

It was getting late. The sun had set behind the hillside to the west of us and the temperature was dropping quickly. In these frigid temperatures, the fuel one gets from food fades fast, especially when you're not on skis or doing something energetic to pump out the body heat. I could feel it first in the numbness of my feet and hands and then in the chills that went up my spine.

Camping on Marble Island would have been the practical thing to do, but Nirlungayuk had no desire to share this hallowed ground of white rock and wooden crosses with the ghosts that haunted it. Like most Inuit, he was superstitious and respectful of a spirit's privacy. Looking down at the snow-covered graveyard below us, I was inclined to agree, even though I did not believe in ghosts. Left with no other choice, we sped back toward the mainland to search for a more suitable place to camp before it got too dark.

This turned out to be more difficult than I had anticipated. Rising and falling tides sent heaps of ice crashing and retreating from the shores of the first two islands we scouted out, making it impossible for us to get our snowmobiles through the jumble of ice built up around the shorelines. An open lead of water to the west of us had also widened dangerously thanks in part to the offshore winds that were now coming in from the east. That eliminated any chance of us taking the shortest route to the mainland.

Without letting me in on his backup plan, Nirlungayuk took off abruptly in a straight line toward the Paschal moon, the first full moon of spring, rising over a spit of land to the south of us. Even with him hauling an aluminum boat, there was no way I was going to be able to go as fast as he was moving. The best I could do was to follow the faint lines of his tracks imprinted on the ice and try not to lose sight of them in the growing dark.

It was bitterly cold. The skin on my face and nose began to freeze, but I couldn't shield my face with one hand lest I lose control of the snowmobile bumping along the ice's rough surface. Nirlungayuk was going so fast that I was beginning to lose sight of him. For the second time that day, I had an irrational and entirely unfounded fear that he might be abandoning me or testing my ability to be on my own for a while.

Whatever fears and doubts I might have had subsided a short time later when I saw the lights of his snowmobile climbing the side of a high hill toward a tiny, unlit cabin. As gloomy and uninviting as the cabin looked from the distance, it at least offered the possibility that we might not have to go through the long, finger-numbing process of setting up a canvas wall tent in the middle of the night in -22°Fahrenheit (-30°C) temperatures and trying to get warm afterward.

Nirlungayuk was already halfway finished digging a tunnel through a 6-foot-tall (2 m tall) snowdrift to the door of the plywood cabin by the time I got there. It was an odd sight seeing him at work; the shadows of two dozen arctic hares he had frightened off looked on in the glow of the moonlight. They were huddled around a ragged red reclining chair that had been discarded on the tundra. "We'll be okay here. I know the woman who owns this place," he said. "I'll fill up her stove with fuel in return for the favor."

It felt colder inside than it was outside the two-room cabin. Nirlungayuk's lighting of a lantern and the two-burner Coleman stove did little to warm things up. Realizing I wasn't going to get any more comfortable watching him trying to light the cabin stove, I got out the broom tucked behind the front door and started sweeping up all the vole and lemming droppings that littered the floor and wooden table.

The cabin was typical of the ones I'd seen all along the eastern and western Arctic coast. The plywood and trim on the outside was painted white and blue. Inside, fake mahogany paneling covered the walls. Several of the white ceiling tiles sagged in spots where rain, melting snow and condensation had leaked through the roof. The one picture on the wall was an enchanting poster of the National Ballet's production of *Snow White*.

In both the front and back room, there were bunk beds with two thin foam mattresses that had been well chewed by the rodents residing here this past and probably previous winters. Cozy it may not have been, but it was preferable to spending the night in a canvas tent. The temperature had dropped below -22°Fahrenheit (-30°C).

Try as Nirlungayuk did, he couldn't get the small stove to ignite.

"I think the fuel line is frozen," he said, pulling his hat off and scratching his head, trying to figure out a way of making it go. "Maybe if I poured some gas in from the outside line, we could thaw it out."

Instinctively, I searched for avenues of escape in the event this strategy resulted in an explosion or fire. Seeing that the stove sat 2 feet (0.5 m) from the front door and 3 feet (1 m) from a window that was too small for my frame, I wasn't exactly put at ease. If I had to flee quickly, I would probably have to sacrifice my hooded down parka to keep me from getting seriously burned.

"So much for global warming," said Nirlungayuk as he stuck a big burning roll of newspaper into the stove as a last resort. A whoosh of flame followed. He snapped his hand back and studied it for a moment to see if any damage had been done.

Soup was a simple fare of frozen arctic char tossed into a boiling pot of melted snow and salt. The hot food, it seemed, bypassed the digestive process and entered directly into the bloodstream. I dipped a rock-hard Pilot biscuit into the broth and savored each bite. Soon, my body was pumping out so much heat that I was forced to take off my parka and my heavy boots.

"My dad sends me char from Pelly Bay," said Nirlungayuk. "I don't know why, but it tastes better than the kind we get here in Rankin."

Nirlungayuk was born forty years ago in Pelly Bay, or Kugaaruk as it is now called, which is nestled along a stark seaside hill of Precambrian rock on the Simpson Peninsula of the central Arctic coast. The tiny community of a few hundred people is as remote as any in the Canadian North. The fact that Nirlungayuk is not from western Hudson Bay is significant in some respects. Even though the Inuit of Nunavut distinguish themselves first as a people, they are very much divided along regional, religious and family lines. A Baffin Islander will stand behind an Inuk from Rankin or Repulse Bay on the national stage defending the polar bear or bowhead whale hunt, but he or she will invariably vote for one of his or her own in matters that are purely territorial or local. These loyalties were at one time so fierce that Protestants and Catholics would barely talk to each other in the same hamlet.

Nirlungayuk learned this lesson early on in life when the Catholic Church sent his father, a catechist, to Iqaluit to convert the Inuit from the predominately Protestant community to the Catholic faith. After a luckless year in which they were largely ignored by the Baffin Islanders, the Nirlungayuks were too homesick to stay on.

The people of Pelly Bay are different from their neighbors in other ways as well. The Aryiligjuarmiut, as they call themselves, are the eastern band of the Netsilingmiut, or "People of the Seal." Unlike the Ihalmiut ("People from Beyond") or the Ahiarmiut ("The Out-of-the-Way Dwellers"), who lived in the barrenlands along the Kazan River and Ennadai Lake and Garry Lake and Back River, they did not suffer the famines that frequently decimated the populations of these other groups. Historically, hunting in Pelly Bay has been extremely good, divided as it has traditionally been among caribou, muskoxen, arctic char and seals. When, for some reason, one harvest failed, the people had another resource upon which they could rely.

Pelly Bay is also unique because it evolved for some time without the overbearing southern influences of the Hudson's Bay Company, the RCMP or competing religious groups that ruled so many other Arctic communities. The Catholic Church, led by Fathers Pierre Henry and Franz Van de Velde, set up a mission there in 1935. Rather than try to assimilate the Inuit into the white man's ways of the world—as many priests and ministers of the time did—the two priests did their best to preserve the Inuit language and way of life.

Nirlungayuk would have liked to raise his family in Pelly Bay—life was good there in many ways. But early on in his teenage years, he could see that the decline of caribou in the Boothia and Simpson peninsulas, the housing shortage that left too many people living in run-down government-built bungalows and the fact that there were so few jobs to be had in a rapidly changing world were going to make it tough on him and his wife, Veronique, to get by with a family as large as the one his mother and father raised.

The route from Pelly Bay to Rankin Inlet was not a simple straight line. First, there was school in Yellowknife and several other jobs that followed. Some were good, others were bad, but all of them opened Nirlungayuk's eyes to a world that was in some ways strange to a boy who was so deeply grounded in the traditional ways of the Netsilingmiut and the tenets of the Catholic Church.

Life in Rankin was also a revelation, given how far removed it is from the traditional world that Nirlungayuk knew. The Inuit here have always lived in and around the area, but the community itself didn't get its start until 1957, when the owners of Rankin Inlet Mine hired local hunters to work on a copper and nickel deposit outside the townsite. It was the first time the Inuit had been employed in a resource venture in the North. The mine shut down, however, after just five years of production.

But the federal government wouldn't give up on trying to assimilate the Inuit into the southern ways of the world. There are people in Rankin today who still cynically recall the time the federal government tried to introduce hog farming into the community after scientists suggested that the caribou populations in the regions might not hold up in the long run. After having underestimated the amount of food a hog can consume, the lords and masters of the project ordered the Inuit to augment the hogs' diet with seal meat. The Inuit caretakers didn't mind the fishy taste of the first hogs that were slaughtered, but the bureaucrats were repelled. Hopeless as the idea was in the first place, it was put out of commission when someone walked into the pens one night and stabbed all the pigs.

Shortly after the hog farm collapsed, yet another federal official came up with the idea of making the Inuit chicken farmers. Whether it was a good idea trying to get walrus and seal hunters to raise eggs for a living turned out to be an irrelevant question. What the architects of the plan failed to appreciate was that the polar bears in the area had other plans for chickens roosting in pens from which they could not escape. By the time the Nirlungayuks showed up, Rankin Inlet residents were more than wise and wary of southern ways.

In spite of these stark differences between Rankin and Pelly Bay, Nirlungayuk and his wife managed to fit in quite well. Not only did she establish herself as a successful artist, they got to go to far-off places through Nirlungayuk's travels on Nunavut Tunngavik Inc. business. For both of their children, Rankin Inlet is now home.

Nirlungayuk realized that the Arctic world was changing in ways that were unfamiliar both to him and his father long before he moved to Rankin. The first hint, he told me, occurred while he working as an airport weather observer at one of the Distant Early Warning (DEW) Line sites on the central Arctic coast. In preparation for the job, Nirlungayuk had to memorize all the cloud formations. The names fascinated him. The formation he anticipated most was the one he had never seen before—cumulus nimbus—the cloud that produces a thunderstorm.

"We just don't get big thunderstorms like you do down south," he said as we sat and drank sweet tea in the dim glow of a kerosene lantern. "But then one day I spotted this cloud formation building up along the floe edge, and it looked like cumulus nimbus. So I reported it to the people at the DEW Line station. No one would believe me. They told me we were too far north for clouds like that to form. I was really frustrated. But when it thundered like crazy later that afternoon, I was a very happy man."

The frost on the walls and ceiling in the cabin were now coming down in drips. After two days of being together, our conversations were also starting to warm up. It was inevitable that the topic would eventually turn to climate change; after all, that was the reason I was there—to better understand the disconnect between scientists and Inuit hunters, who don't always agree on how it is affecting the Inuit and the Arctic world in which they live. Nirlungayuk seemed eager to share his thoughts with an outsider like me, who was neither a scientist nor an Inuk.

"I respect Ian Stirling," he said of the veteran polar bear scientist. "He is a good scientist and a good man. He says polar bears are declining. That's fine. He's been doing this kind of research for a very long time, but how do I respond to people like Raymond and other hunters who insist there are more polar bears out there than they've ever seen? You talked to Raymond last summer. He said there were so many bears around his community in Coral Harbour that it was dangerous to go out at night."

This was true. Elder Raymond Ningeocheak, Nunavut Tunngavik Inc.'s longtime second vice-president, was absolutely certain that what he and others were seeing in the Arctic was a more accurate reflection of what was happening on the ice. But there were others at the Coastal Zone conference in Tuktoyaktuk who expressed fears that climate change would bring an end to the Inuit way of life. Inuvialuit hunters from Sachs Harbour and Holman (Uluhaktok) described how climate change had reduced access to harvesting grounds and altered the migrations of some species. They puzzled over Pacific salmon that were showing up in their nets and grizzly bears that were crossing over from the mainland and breeding with polar bears. Some birds, such as the robin, were so new to them that they had no word for them in their language. John Keogak, an Inuit hunter from Sachs Harbour on Banks Island, was among the most vocal. It was his view that the Inuvialuit and the Inuit needed all the help they could get from scientists and the government to adapt to this unfolding new world. His one big complaint was that scientists would often come up to do their research but would rarely come back to tell them what they found.

Wide as the gap is sometimes between science and traditional knowledge, there are examples in which the two contrasting views of the natural world have blended together positively. Scientists now know that the carnivorous rogue walrus, such as the one that had nearly taken the life of that hunter earlier on our trip, is not merely the stuff of Inuit myth. It is real and potentially more dangerous than its more sociable clam-eating cousin. Thanks to the Inuit, scientists know that the harlequin duck, not seen for more than seventy years on Baffin Island, is still breeding there and that ivory gulls in the High Arctic are declining dramatically. They've also learned from the Inuvialuit and Gwich'in in the western Arctic where Dolly Varden fish populations are at risk and the health of seals seem to be in decline.

Social scientists and Inuit leaders have justifiably made much of these insights in recent years. Academics are now routinely churning out papers on the value of traditional knowledge (TK). Many of them talk about TK in terms of epistemology. The Nunavut Social Development Council recently refused to use the term *traditional knowledge* because of the suggestion that its relevance was of the past, not the future. The council insists on using the term *Inuit Qaujimajatuqangit*, or IQ, which they say includes all aspects of traditional Inuit culture, be it language, social organization, knowledge, life skills, perceptions and expectations.

What few people in the academic community talk about in any meaningful way is that this "way of knowing" is sometimes influenced by things that have nothing to do with being Inuit in the traditional sense. Evangelical groups have convinced so many Inuit communities about the wisdom of their way of knowing that some teachers in northern Quebec have been told to stop teaching evolution to Inuit children. The creationists' point of view is now taking hold in some Nunavut communities where Glad Tiding ministries have significant numbers of people talking in tongues on Sundays. How evangelical teachings, television and southern movies shape the Inuit "way of knowing" is rarely discussed.

In their zeal to promote this "way of knowing," many TK and IQ proponents have also overlooked the possibility that science is useful in pointing out how contaminants are entering the bloodstream and breast milk of Inuit mothers and how diseases, never seen before, have infected some marine mammals. There is also no one discussing the possibility that the Inuit themselves are not infallible—that falling through the ice, for example, may have resulted from risk-taking or youthful inexperience rather than a climate-changing event.

Grant Gilchrist and Mark Mallory, the scientists who benefited from traditional ecological knowledge when they were studying ivory gulls and harlequin ducks in the eastern Arctic, learned in another innocent way that the Inuit are fallible when they went to Greenland to talk to colleagues about local knowledge of thick-billed murres in the Upernavik region. The seabird lives year-round in the Arctic, nests on exposed cliff edges and often migrates from the Greenland side of Davis Strait to Baffin Island in Canada. Since 1965, when murres were first harvested for commercial purposes in Greenland, the numbers have declined dramatically. None of the Greenlanders interviewed disputed this fact. But when it was suggested that overhunting might have been part of the problem, most of them were in denial. They were convinced that the shootings at the colonies had merely "upset the birds," prompting them to move elsewhere. Search as biologists in Greenland did, they could not find any spot on the Greenland coast where the birds might have gone.

Faced with similar harvesting issues exacerbated by climate change as well as overhunting, many Inuit hunters in Canada have also been in denial about steep declines in Peary and barren ground caribou populations. In the case of Peary caribou, they have steadfastly refused to reduce their hunts or shift their focus to the more plentiful muskox populations. Nor will they allow scientists to set up a captive breeding program that could aid in a future recovery of Peary caribou, or at least preserve the genetic stock of this quintessentially Canadian animal. All attempts by the Committee on the Status of Endangered Wildlife in Canada to relist the Peary caribou as endangered have failed because the Inuit have so far been successful in arguing that they were not properly consulted.

Nothing illustrates how serious the standoff between science and traditional knowledge has become than the ongoing debate over polar bears in western Hudson Bay. Long-term research by Ian Stirling and other scientists indicates that climate change is hitting this population hard. With less time to hunt seals on rapidly receding sea ice, polar bears are getting thinner and producing fewer cubs that survive the first year of life. Between 1987 and 2004, the population has declined by about 22 percent. Stirling insists that the data is as clear as he and his colleagues will ever get on any polar bear population.

Up until this point, many Inuit hunters—not all—have shown little interest in reducing their annual quotas. They continue to insist that Inuit *Qaujimajatuqangit* tells them there are more animals than ever. They are not alone. In a show of solidarity, Nunavut environment minister Patterk Netser, an Inuk himself, took that message to Washington in the spring of 2006 at hearings into the U.S. Fish and Wildlife Service's proposal to list the polar bear as "threatened" under the Endangered Species Act. In pleading his case, Netser described fears of climate change threats to the animals as "hysterical." He had harsh words as well for those who claim polar bears are coming into communities because they're starving or because ice conditions have changed their patterns of migrations.

"The suggestion that they are so easily misled is not only silly, but also shows a disturbing lack of respect for indigenous knowledge," Netser said.

Raymond Ningeocheak further inflamed passions a few months later when he took issue with scientist Andrew Derocher for suggesting the possibility that the apparent increase in polar bears cannibalizing one another in the Beaufort Sea region might be a sign of climate change. He denied that melting sea ice was undermining the bears' ability to find seals. In a letter to the editor of *Nunatsiaq News*, the newspaper of the eastern Arctic, Ningeocheak countered with the claim that it was just as possible that there was an overabundance of bears hunting for a limited amount of food.

Not every Inuit leader shares Ningeocheak's sentiments. Sheila Watt-Cloutier coined the phrase *right to be cold* when she filed a petition on behalf of sixty-two hunters, elders and women from Canada and Alaska to the Inter-American Commission on Human Rights in December 2005. She was requesting relief from human rights violations due to climate change.

At the time, Watt-Cloutier was chair of the Inuit Circumpolar Conference, the international organization that represents Inuit from around the polar world, and a candidate for the Nobel Prize. Her main goal back then, as it still is now, was to put a human face on the climate change issue. For a hunting culture, she said, climate change is a question of survival. "Our human rights—to live our traditional way—are being violated by human-induced climate change."

Watt-Cloutier sympathized with the plight of polar bear hunters when she filed the petition. But she expressed her fear that the emotional debate over polar bear quotas would deflect attention away from the bigger threats that climate change poses to Inuit culture. The Inuit will lose, and lose big, she warned, if there is no meaningful reconciliation between proponents of traditional knowledge and scientists conducting research.

Just how such reconciliation will come about is far from clear. So far, both the federal and territorial governments have behaved as if polar bears, beluga, narwhal and other Arctic animals are solely owned by the Inuit. It's become increasingly common to deny funding for a research project, a recovery program or a survey simply because Inuit hunters in a community object to it. Increasingly, there's less recognition that responsibility for polar bears, migratory birds, endangered wildlife and other species should be a national and international priority. Since the species-at-risk legislation came into effect in Canada in 2003, the scientific committee that oversees the program has recommended the listing of five animals that live in the Arctic. To date, the Inuit have successfully prevented adding any of those animals to the list. The Canadian government has also been more than happy to devolve responsibility for wildlife to Nunavut and the Northwest Territories, but it hasn't provided the financially starved territories with the resources required to manage wildlife effectively. Many Inuit, on the other hand, are brought along on research projects simply as spectators, not as participants, or as polar bear monitors in regions where there are rarely polar bears.

This has not always been the case, and it is changing in some instances. In the western Arctic, scientists such as Ian Stirling and Tom Smith forged strong working relationships with the Inuvialuit more than twenty-five years ago by partnering up with hunters such as Jimmy Memorana of Holman (Uluhaktok), just as Jack Orr is doing in his whale capturing efforts. Dick Harington and William Irving of the Canadian Museum of Nature did the same thing with the Gwich'in of Old Crow in the northern Yukon and were rewarded handsomely for the overture. Harington's discovery of woolly mammoth remains on the Whitestone River in the Yukon would not have occurred had he not taken seriously the Gwich'in elders who told him about enormous creatures that once lived in that part of the Yukon.

Since archeologist Charles Schweger picked up where Harington and Irving left off, the marriage between science and traditional knowledge has led to a number of other insights in the Old Crow region that would be too hard to believe if they couldn't be confirmed. In 2007, there were as many as fifteen scientists collaborating with the people of Old Crow in some way. The research is focusing on everything from monitoring water levels and wildlife populations in the Old Crow Flats to the stability of permafrost in the region.

Lois Harwood has taken this collaboration one step further. Several years ago, the Fisheries and Oceans scientist did not have enough money or time to collect all the samples she needed to determine the health of whales and seals in the Beaufort Sea. So rather than give up, she hired Frank Pokiak, who had greeted me warmly in Inuvik, and John Alikamik, Inuvialuit hunters from Tuktoyaktuk and Holman (Uluhaktok) respectively, to collect and package organ samples at their camps.

Pokiak sees this as win-win situation. By being involved in the science, he told me when I met up with him a year earlier, the hunters get a better scientific explanation for the changes they know are happening. Their involvement also assures the Inuit, Inuvialuit and Gwich'in that the animals they're hunting for are well managed.

I couldn't pretend that I was well rested when I got up early the next morning. Visions of our little stove springing a toxic leak or a cabin-burning flare-up haunted me all night long. When I finally managed to nod off, I woke up in a sweat, having failed to anticipate that the stove would eventually produce too much heat to allow me to stay zipped up in a sleeping bag that was rated for -31°Fahrenheit (-35°C). I was grateful, however, for the pot of coffee already percolating on the stove and the fact that I didn't have to put on my parka to eat the fish soup that Nirlungayuk had reheated.

Outside, the sun had not yet burned a hole through the ice fog that had descended in the early hours of the morning. Ice fogs occur when water droplets in the air are frozen into crystals. This usually happens when the temperatures fall below -31°Fahrenheit (-35°C). Ten minutes after I hung my thermometer out the door, it was showing -34.6°Fahrenheit (-37°C), which is not common but not unheard of at this time of year in western Hudson Bay.

In no hurry to venture back onto the ice, where we wouldn't be able to see what lay 10 feet (3 m) ahead of us, Nirlungayuk decided instead that it would be better to sit and drink another pot or two of coffee. That's when I discovered why he was always going so fast on his snowmobile. It had nothing to do with making a fool out of a *Kabloona* (non-Inuit) like me. It had more to do with his passion for NASCAR racing. "I'm not sure why I like it so much," he shrugged. "Maybe it's the sound of the engines roaring that reminds me of going fast on my snowmobile. Since I was a boy, I loved that sound. I love going fast."

Nirlungayuk was neither a big nor a small man by any measure, but he was built like a rock. When he looks at you, it seems as if he's searching for signs that you really understand what he is saying. This, I imagined, evolved naturally in a job in which he has to explain to people certain "ways of knowing" that might not be obvious to those who live down south. It is one thing to be traveling in ice fog and quite another to describe it to someone who has never experienced the surreal feeling of being in an icy tomb such as this. No amount of words can adequately explain how limited vision, the total lack of smell, the stinging sensation on the skin and the amplification of sounds are all part of the weird experience.

There is also the complicating factor of being Inuit. Being Inuit means that one is "of the land" or of the creatures that inhabit the land. That is what the "miut" in "Netsilingmiut" ("People of the Seal") and "Ihalmiut" ("People from Beyond) means. It is a difficult concept to explain to someone from Toronto or New York whose firsthand understanding of nature comes from visits to High Park or Central Park.

When the sunlight finally leaked into the sky later that morning, we headed back to the sea in search of walrus. Nirlungayuk was certain that they were trapped somewhere beyond the vast ice sheet that extended as far as we could see. He wasn't intimidated by what lay ahead, but I was. I couldn't help wondering how James Knight and his men felt that second spring when they looked north and south and saw nothing but cold blue skies and the cold white expanse of ice and snow. Life was absent here. The isolation they endured offered no escape. Without a ship and provisions, it must have been too terrifying to leave the island, prison that it was.

In the end, we never did get the walrus Nirlungayuk was hoping for, which was fine with me. I had trouble envisioning the two of us hauling 800 pounds (360 kg) of blubbery flesh out of the icy sea. Nor did we see the Inuit families I thought we'd meet while traveling along the floe edge.

Perhaps it was because spring was so late in coming this year. But twenty years ago when I was on a similar Inuit hunting trip in Hudson Bay, there were tent camps lined all along this floe edge. The springtime event was similar to fall harvests and Thanksgiving get-togethers down south. Once we got back to Rankin Inlet, I figured out what everyone in Rankin had been up to while we were away. Most of the town's young people were drag-racing on the sea ice, doing their best to pop the pistons in the engines of their snowmobiles.

Later that night, I had dinner with Nirlungayuk and his wife, Veronique, before heading back to Nanuq Lodge, the bed-and-breakfast that John Hickes owns and runs with his partner, Page Burt, a noted Arctic naturalist. John was born sixty-two years ago at a remote outpost at Pistol Bay near Whale Cove on the west coast of Hudson Bay. As the son of a white father and Inuit mother, he has roots in both cultures. More than anyone else, he can see things from both sides. John was one of the original "whale riders" of Churchill, Manitoba, a small group of Inuit men who helped scientists find a way of capturing beluga whales in the Nelson River. He is also a former mayor of Rankin Inlet, a motel and restaurant owner and president of Nunasi Corporation for eight years. Nunasi is the investment arm of the Inuit land claims organization. Under Hickes's leadership, it became a major shareholder in Canadian North Airlines and other northern businesses.

At the time, Hickes wasn't all that happy about the direction that Inuit culture was going. "A while back, there was a photo in the newspaper where this kid was sitting on top of a polar bear he had just killed.

He couldn't have been more than twelve years old. It wasn't his age that bugged me. What bugged me was that he was sitting on the polar bear with earphones listening to an iPod. Don't tell me that's the Inuit way. That's not the way we show respect when we kill an animal."

Hickes sees much of what is going on as growing pains that can be corrected if the Inuit are prepared to embrace the future as well as the past in a meaningful way. "It's really a question of respect for the past and recognizing that the future is also about change. You can't always expect to succeed if you're going to do things the old way. If we did, we wouldn't be owners of an airline, and all that money we got from the land claims would be worth a lot less now than it is."

Before going to bed that night, I went to the common room to make myself a cup of tea. While waiting for the water to boil, I listened in on a conversation between two Inuit men and an Ottawa-based scientist with the Canadian Nuclear Safety Commission, who was going to be giving a talk on a uranium mining development proposal east of Rankin the next day. In a sincere, but somewhat arrogant manner, the young scientist from Ottawa was telling them why nuclear power is not really the environmental threat that it used to be. Both Inuit men listened politely but didn't say a word.

I held my tongue when the scientist kept referring to Nunavut as Nunavik, as if he had no idea where he was. After the third or fourth time that he repeated this mistake, Mike Ilnik, an Inuk businessman who had served in the Canadian Forces, finally snapped.

"You people are always coming up here and telling us what to do, and you can't even get the name of our territory right," he said exasperatedly. "Nunavik is northern Quebec, this is Nunavut. It's the same with Iqaluit. The *q* is not a *k*. It makes an *h* kind of sound."

Moments later when I introduced myself to Ilnik, I realized by the smile on his face that his rant was a lot more good-natured than it sounded at the time. "I know the kid meant well," he said. "But when you're in the Forces as I was, you learn to speak up when you have to."

After several minutes discussing what life was like now that he was back home in the Arctic, I asked Ilnik what he thought about climate change and the impact on Inuit culture. He barely paused to give me an answer.

"Booze and drugs is going to ruin our culture long before climate change will," he said. Then he cocked his head toward the nuclear safety scientist who had just left the room. "That uranium mine he's talking about is going to be right in between two caribou calving grounds near the Thelon Game Sanctuary. We've got more than enough things to worry about these days."

Italian restaurateur Joe Grande, second from right, teaches the Inuit narwhal-capture team from Repulse Bay how to play bocce.

Hunters from Repulse Bay carve up the carcasses of several narwhal on a small ice floe at the very north end of Hudson Bay.

A wooden cross marks the grave site of one of the men who died on James Knight's ill-fated expedition in search of gold and a Northwest Passage in 1719–1721.

Like many Inuit hunters, Gabriel Nirlungayuk believes that scientists and policy makers do not fully appreciate the wisdom that comes from the traditional knowledge of elders.

Deteriorating ice conditions in Hudson Bay and other regions of the Arctic are making it increasingly dangerous for the Inuit to hunt.

Virtually absent from Banks Island in the High Arctic for more than a century, muskoxen have made a dramatic comeback in recent decades. More than 60,000 animals, more than half of the world's population, now reside on this one island.

The Peary caribou is an animal unique to Canada. The High Arctic population has dropped by more than 80 percent since the early 1960s.

Most of the barren ground caribou herds in the Arctic are in a freefall. No one knows why, but climate change, resource development, hunting pressures and others factors are thought to be responsible.

The red squirrel is one of the few species that has taken advantage of climate change in the North.

Scientist Scott Dallimore of the Geological Survey of Canada looks on at the Mallik well site at the north end of the Mackenzie Delta. He and an international team of scientists are trying to find a commercially viable way of extracting natural gas from methane hydrates, which have the potential to be the next great fossil fuel source for an energy-starved world.

One of the wolves that regularly visited the author as he explored the north end of Ellesmere Island.

A scientist looks on at one of a number of glaciers that slide out of the icefields of northern Ellesmere Island.

A polar bear walks along the shores of Hudson Bay in fall, waiting for the ice to form. More than anywhere else, these bears are spending more time fasting on land.

chapter eight

CARIBOU CRASH

– Cumberland Sound, Baffin Island –

What the buffalo was to the North American Indians in days gone by, the (caribou) is now to the Eskimos and other natives of the north country.

—James W. Tyrell, *Across the Sub-Arctics of Canada*, 1908

We were halfway up Cumberland Sound along the southern coast of Baffin Island keeping an eye out for big bowhead whales when Robbin Dialla turned and pointed to a bull caribou that was coming down the side of a huge rockfall. We had just spent ten days with Jack Orr from the Department of Fisheries and Oceans capturing beluga whales on another one of his projects. There was no need to augment our food supply. But Dialla couldn't resist the opportunity to take fresh meat home. He landed the boat, loaded up his gun and shot the animal.

An hour later, the caribou, which had scarcely been aware of our presence until the first of three shots was fired, had been gutted, skinned, carved into chunks and piled into the boat. It happened so fast and so effortlessly that I barely noticed how much blood had soaked into my survival suit while I helped load up. Neither did the other Inuit hunters who shared the boat with me back to the community of Pangnirtung. The scene had played out so many times in their lives that the sight of blood on clothes was probably much too common to deserve notice.

Although the caribou is not as iconic an Arctic animal as the polar bear, it is more critical to the welfare of the Inuit and to the Dene, Gwich'in and Métis. In most communities, the measure of a hunter is judged by his skill in killing enough animals to feed not only his family, but other members of the community who may not be healthy or old enough to go out on the land. A single caribou saves a family $500 to $1,500 that they might otherwise have to spend on store-bought meat. On a broader scale, the value is magnified greatly. In 2008, a report produced by InterGroup Consultants pegged the total economic value of the Beverly-Qamanirjuaq, two herds shared by Métis and Aboriginal peoples from Nunavut, the Northwest Territories, northern Saskatchewan and northern Manitoba at $20 million annually.

The value of caribou in Aboriginal culture extends far beyond economies of scale and the measure of a hunter. Wherever you go in the Arctic, the caribou is found in the clothes that people wear, the prints that artists draw, the soapstone they carve and in the myths and legends that people tell.

One legend suggests that caribou came to Earth specifically for the well-being of Aboriginals. This story has many versions, but one that was told to Knud Rasmussen, the Danish ethnologist who was mothered by a Greenlandic Inuit woman, conveys a universal theme. Rasmussen was traveling across the North American Arctic between 1921 and 1924 when a man by the name of Kibbarjuk told him of hard times, when there were no more caribou in the world. According to Kibbarjuk, it was left to the imagination and hard work of a single man to come up with the idea of digging a hole in the Earth and allowing the animals to rise to the surface in great numbers. When there was enough caribou to feed everyone in the world, the man filled the hole back in.

The peoples of the Arctic have come a long way since the days when shamans were called upon to frighten off spirits thought to be responsible for the famines, which inevitably followed when caribou veered off the beaten track. Most everyone would now concede that once the caribou are gone, there will not be a hole from which they'll magically return.

Twenty years ago, the possibility that caribou might vanish from the North was inconceivable. There were several million caribou in Alaska, Greenland, the Northwest Territories, northern Quebec, Labrador and what is now Nunavut. Since the 1990s, however, many of these great caribou herds have been in a free fall. The Bathurst herd in the central Arctic numbered 472,000 in 1986. When wildlife biologists did a census in 2003, the herd was down to 128,000 animals. Farther west in the Northwest Territories, the Cape Bathurst herd had 17,500 animals in 1992. Recent estimates suggest there are no more than 1,800 animals today. The Porcupine caribou herd, which is found in Alaska, the Yukon and Northwest Territories, declined from 178,000 animals in 1989 to 123,000 in 2001. Some scientists speculate that the numbers could be much lower now, although weather on the north slope of Alaska and the Yukon hasn't allowed them to confirm their suspicions. At an international gathering of Gwich'in peoples in Old Crow in the northern Yukon in the summer of 2008, caribou hunters from both sides of the border were presented with a computer model that considered what a worst-case scenario looked like. The model pegged the population at just 90,000, which would represent a drop of nearly 50 percent in two decades.

Some of the smaller herds, including the Peary caribou in the High Arctic, are doing so poorly that they have all but disappeared. Even the reindeer herd that was brought over from Alaska in the 1930s has pretty much vanished from its small corner in the Mackenzie Delta. Reindeer are caribou that have been domesticated for agricultural purposes. The Alaskan animals were intended to feed Inuvialuit who had fallen on hard times after the whalers cleared out and left many of the wildlife resources in the western Arctic severely diminished.

In the million or more years that caribou have roamed the circumpolar world, there have been countless ebbs and flows in the populations.

Don Thomas, a Canadian Wildlife Service scientist who searched the records to get a profile of caribou populations over the past century, suggests the low numbers of caribou seen during the 1920s was followed by a rebound in the 1940s. The numbers steadily declined from 1950 to 1970 before peaking in the 1980s and dropping hard around 1990 through to the turn of this century.

In the late 1970s, there were so few animals remaining in the Beverly and Qamanirjuaq herds on the barrenlands of what is now west-central Nunavut and the western end of the Northwest Territories that scientists were urging the Inuit to stop hunting them. Those same scientists then watched with amazement and some disbelief as the numbers on the Qamanirjuaq range went from a low of 39,000 in 1980 to 230,000 in 1983, 260,000 in 1987 and then 496,000 the last time the animals were counted in 1994.

Part of the problem back then, as it is to a lesser extent now that aerial survey techniques have improved, is finding enough animals to get an accurate estimate of how many caribou are on the range. Even in a year when weather and the migratory animals cooperate with the scientists, a good count has a possibility of error in the range of plus or minus 10 percent. In less favorable circumstances, that margin of error can be as high as 20 percent.

The huge size of the range and the enormous numbers of animals make it difficult and extremely expensive to get a more accurate measurement. If the location of migration routes and calving ground sites shift just before or during a census, as they can if spring snow conditions warrant, the margins of error may be even higher. It can be maddening for a scientist who has to explain these variables to the public and to Aboriginal leaders who wonder why the numbers are so volatile.

No one knows what's behind the most recent declines in caribou populations. The best guess is that climate change, natural variability that comes with decadal shifts in atmospheric regimes over the north Atlantic and the Arctic, overhunting and resource development are all likely playing a role. Industrial contaminants that are being transported into the Arctic and creeping into the animals' food supply could also be a factor.

This was an especially big concern in the 1980s when the radioactive fallout from the Chernobyl nuclear disaster in Ukraine began showing up in caribou in the North American Arctic.

Scientists have come to realize that the biggest reason for these ebbs and flows lies in a complex relationship between the caribou's ability to find food and its ability not to be food. Invariably this relationship is set to what Anne Gunn, a caribou scientist with more than thirty years of experience in the field, describes as "the rhythm of weather and climate." Simply put, long stretches of cold, dry winters with less snow tend to favor caribou because there is little to slow them down while they're on the move or being chased by wolves. Less-dense and hard-packed snow also makes it easier for the caribou to dig down to the vegetation they need in order to survive.

Stretches of warm, wet winters, on the other hand, can be brutal for opposite reasons. Not only is there a possibility that the snow will be deep during the long migration to the calving grounds, but thawing and melting can cause some of it to ice over. If those winters are followed by hot, dry summers that favor parasites, biting flies and fires that destroy nutritious lichen, the results can be catastrophic.

The Peary caribou, a diminutive animal that is superbly adapted to living in some of the most severe climatic conditions in the northern hemisphere, are extremely vulnerable to the warming already beginning to change the rhythm of weather and climate on the High Arctic Islands. So, in some cases, are muskoxen—shaggy, prehistoric-looking animals that survived the last Great Ice Age.

Frank Miller learned this in rather dramatic fashion in 1996 when he spied a dark spot on the sea ice while flying along the coast of Bathurst Island in the High Arctic searching for Peary caribou. Assuming it might be a small herd migrating to a nearby island, the Canadian Wildlife Service scientist instructed the pilot to go in for a closer look. Seconds later, he could see that these animals were not the caribou he was looking for, but muskoxen.

Muskoxen often form a circle when threatened, positioning their rumps inward so that they can get a 360-degree view of the wolf pack or the human hunters that surround them. The animals in this circle didn't bolt as muskoxen normally do when an aircraft hovers above them. Puzzled by this behavior, Miller got the pilot to land a few hundred feet away. Even as he approached on foot, none of the animals flinched. Miller walked toward them. Only when he got to within a few feet did it dawn on him: they were all dead. The animals were frozen stiff, leaning against one another like statues that had been knocked over by the wind.

It was one of the strangest things Miller had seen in his thirty years of fieldwork in the Arctic. He guessed that the animals were probably on their last legs and starving when they headed out across the sea ice searching for better food conditions on another island. After they sapped what little energy they had left trudging through the deep snow, the muskoxen tried to dig down to see if they could find food. Discovering there was nothing there but sea ice, the animals huddled in that defensive circle, and then gave up. The snow melted and eventually hardened around their bellies, propping them upright until they died.

This gruesome sight did not bode well for Miller's search for Peary caribou. In the weeks that followed, Miller saw plenty of the succulent saxifrage flowers that Peary caribou feed on in summer, but precious few animals and not a single calf, as would be expected at that time of year. What he did discover with alarming regularity were the carcasses of both caribou and muskoxen strewn across the tundra. By the time Miller completed his study that summer, he had counted just three hundred live caribou, a small fraction of what he had expected to see. When the die-off finally ended two years later, almost 98 percent of the caribou on the south-central Queen Elizabeth Islands three years earlier were gone.

This wasn't the first time Miller witnessed a catastrophic die-off. In 1974, his third season in the High Arctic, he was concerned that he may have missed something when he found only a tiny fraction of the four thousand caribou that scientist John Tener had counted in the same area of the Queen Elizabeth Islands thirteen years earlier. Only when he went back home to Edmonton and searched through the meteorological records for clues to what might have happened did he come up with a plausible explanation. Those records indicated that freezing rain had fallen over the area in fall of 1973. This was followed by heavy snow in winter and recurring periods of thawing and freezing the following spring. As a result, much of the Queen Elizabeth Islands had been transformed into a giant, snow-covered skating rink. The ice was likely so thick in most places the animals couldn't get through to the vegetation. Those that were successful probably spent more energy than they received.

The Queen Elizabeth Islands are among the most remote and difficult places to get to in North America. Stark and extremely inhospitable, some of the islands are little more than giant pancakes of sand and gravel a few feet above the sea. Because they are covered in snow and ice, or shrouded in fog for almost twelve months of the year, nothing much grows on them. But even this far north, there are oases of life in places such as Polar Bear Pass on Bathurst Island, at Dundas Peninsula on Melville Island and at Walker River, where Miller spent more than two decades studying the region's Peary caribou population.

Because Peary caribou live in such remote locations, it was quite some time before anyone got round to studying them in any systematic way. In the 1960s, when John Tener did that first survey, he estimated that there were forty-nine thousand Peary caribou in the Arctic, twenty-six thousand of them on Bathurst and other islands in the Queen Elizabeth archipelago. No one followed up on that first count until Miller came around in the 1970s and estimated that 85 percent of the caribou and 70 percent of the muskoxen had disappeared from the Queen Elizabeth Islands as a result of that widespread icing of the tundra.

Although the muskoxen population has since rebounded, the downward slide of Peary caribou has not, at least not in any appreciable way. By the turn of the century, the total number had dipped to below eight thousand. The High Arctic population was in such deep trouble by 1997 that the Canadian military was brought in that fall to relocate some animals to a Calgary zoo site in the hopes of preserving their genes or possibly bringing them back in the future. A giant Hercules airplane got as far as Resolute on Cornwallis Island before a fall blizzard forced the crew and a wildlife capture team to turn back.

The military's return is as unlikely as is a recovery any time soon. The Inuit have made it known that that they are philosophically opposed to the relocation of any animal in the Arctic and they are resolute in their determination not to have the Peary caribou listed as endangered. Peary caribou also did not fare well in the 1990s, and the very few surveys done in recent years suggest only a nominal increase in numbers on Ellesmere and Banks islands. What's more, no one except Debbie Jenkins, a Nunavut biologist who works in the eastern Arctic, and John Nagy, a Northwest Territories biologist working in the west, has picked up where Miller, now retired, left off. With many other responsibilities, both of those biologists have comparatively little time to devote to the animal.

Shortly after that trip to Cumberland Sound, I spoke with Frank Miller in Edmonton. He had lost some of the bulk that made him such an intimidating figure when I had first joined him at his base camp in the High Arctic ten years earlier, but the former U.S. Marine still growls when he talks and he doesn't suffer fools gladly. Climate change didn't kill the Peary caribou in the 1970s and probably not in the 1990s, he insisted. Catastrophic die-offs like this, he noted, have been occurring for thousands of years. Invariably the caribou populations bounce back.

What's different now is that there are more people hunting caribou, more resource exploration displacing caribou habitat and in all probability more wolves killing what few animals are left. If a warmer future increases the number of these freezing and thawing episodes in the High Arctic, Miller warns, the era of the Peary caribou will be over very soon.

Of all the characters in the colorful world of Arctic science, Miller seems more suited to the study of polar bears than the diminutive Peary caribou. Polar bears, in fact, were the focus of his research when he left the Marines and his home in Connecticut for a wildlife biologist position in Canada in the 1960s. Six months into the job, it became abundantly clear that funding for polar bear research was in short supply, so when the Canadian Wildlife Service embarked on an ambitious study of the Qamanirjuaq caribou herd in the barren lands in the late 1960s, he volunteered to be part of the research team.

Making the change from the ever-dangerous polar bear to the shy and skittish caribou did not make the job any less perilous in those early days. Instead of using tranquilizing dart guns, Miller and his colleagues would use shepherd's crooks and 22-foot (7 m) freighter canoes to haul in and tag caribou on stormy Arctic lakes. The switch to caribou turned out to be a good fit. During three decades of study, Miller produced more than two hundred scientific papers, which made him a world authority on muskoxen, Peary and other caribou.

Much as Miller would like to see something done to save the Peary and other caribou populations in the Arctic, he believes there is a dangerous misconception that moving animals from one population to another on the archipelago and, even worse, from the mainland to the archipelago, would be a valid management option. This is a strategy that is gaining momentum in some circles.

Miller doesn't think such a move should be undertaken. "Captured caribou moved by people between or among geographic populations would change the existing genetic differences between the caribou," he told me. "This would effectively destroy thousands of years of evolution of separate evolutionary lines of caribou on the Canadian Arctic Archipelago."

Emergency or supplementary feeding, artificial insemination or embryo transplant schemes might be useful in implementing a small-scale recovery plan. Miller believes the costs of such activities, the difficulty of carrying them out and their unknown value make them relatively small-scale, last-ditch efforts at best.

Most everyone now believes that a captive breeding program such as the one that was aborted at the last minute in 1997 would likely be unsuccessful in restocking the High Arctic in the future because there is little evidence to suggest that semi-tame animals raised in captivity could be returned to such an extreme environment. There is also the potential for these animals to become carriers of diseases that would seriously harm whatever animals might be left in the Arctic.

Miller still believes that presence of caribou at the Calgary Zoo would go a long way in educating people about the plight of a uniquely Canadian animal that decision-makers have all but abandoned.

"It's most unfortunate that the captive breeding proposal was rejected by the people of Resolute Bay and Grise Fiord," he told me. "If Peary caribou from Bathurst Island or Melville Island had been captured at that time and placed in the facilities at the Calgary Zoo under the watchful eyes of experienced veterinarians, there would now be a ten-year increase in those Peary caribou, most of which would already be breeding and producing their own young at the zoo."

What Miller would like to see is the Inuit shifting their harvest strategies to muskoxen, which, with few exceptions, have not been affected in any serious way by hunting, climate change and resource development so far.

"The chances for the recovery of a caribou population are much higher with no hunting of those caribou," he said. "There are enough muskoxen around to meet the requirements of every settlement dependent on Peary caribou and all other Arctic Island caribou throughout the archipelago. The switch to muskoxen would promote the conservation of a caribou population without causing any lasting harm to the harvested muskox population, which could be harvested within sustainable limits."

Miller acknowledges that because of the Inuit's strong cultural interest in caribou hunting and because the eating of caribou meat is as much a means of socialization as a source of food they favor, it will be difficult for them to make the switch. He feels that the alternative—a world in which caribou are too scarce to be a regular item on the dinner menu—would be tragic. "We take it for granted that caribou will be around forever because there have been so many for so long. But if the history of wildlife management tells us anything, it tells us that wildlife species, especially those that are limited in number across a very narrow geographic range, can disappear very quickly."

The prospects for barren ground caribou populations on the mainland of the Arctic are similar and yet fundamentally different from those of the

Peary caribou. Unlike Peary caribou, which have a relatively small range of movement in the High Arctic, the larger barren ground animals travel thousands of miles each year to get to the calving grounds on the coastal areas of the Arctic. In some cases, the size of the herds are in the range of a half-million animals. Predicting when these animals will move and what routes they will take has been one of the great challenges for scientists trying to count them.

No one knows this better than Don Russell, a Canadian scientist who has studied the Porcupine caribou herd in the Yukon and Alaska for more than twenty-five years. I met Russell in the field in 1981 when I joined him on an aerial survey to monitor the spring migration of these animals. As it turned out, we were about a week ahead of the animals that year and ended up seeing more moose than caribou along the coast.

When Russell and I met again in the summer of 2008 in the Yukon, he confessed that the animals continued to surprise him by showing up in places where they were not expected.

Russell is short, stocky and bearded. Most people call him Donny, and not just because his middle initial is *E*. In a world in which there is so much at stake and in which the science of caribou biology is constantly under attack from within and without, Russell's easygoing nature, longevity, research and insights have made him one of the more popular figures in the world of Arctic science.

Russell's trip north is a little different from the paths of most northern biologists who come north straight out of university. He was working in northern British Columbia in 1976 when a friend passed on word that the Yukon government was looking for a habitat biologist to fill out its newly minted wildlife department. Having done graduate work on caribou at Prudhoe Bay in Alaska, Russell was more than qualified for the job. The number of people who had field experience in the Arctic at the time was limited at best.

Although research budgets tended to be lean in those early years, the job went well. The department grew so fast, however, that Russell eventually ended up doing more managing than working in the field. When the

Canadian Wildlife Service recruited him for a senior caribou research job in Whitehorse, he jumped at that chance to be one of just two people in the Yukon office that was managed from Vancouver. "I liked the idea that my boss was 2,000 miles away," he said.

Since then, Russell and his colleagues in the Yukon and Alaska have put together a profile of the Porcupine herd that covers everything from calving rates to predator profiles. They now know how caribou respond to insect harassment, weather, snow depths, hunting and changes in vegetation. It's as comprehensive as any long-term study of a large mammal in the Arctic as there is.

The Porcupine herd is named after the big river that cuts through its range, which is about half the size of the Yukon, or the entire state of Wyoming. Traditionally, the animals spend the fall and winter in close proximity to the tree line in the interior of Alaska and the Yukon. Sometime in April, pregnant females begin a 300-mile to 400-mile (500–700 km) journey to the calving grounds on the coastal plains where there are fewer biting insects and predators, and plenty of sedges, grasses and lichen for them to eat.

By early June, most of the cows will have reached the coast. The rest of the herd will have caught up by the end of the month. The animals bunch up in the thousands and sometimes tens of thousands when the heat of the summer produces hordes of mosquitoes and biting flies that can literally drive some of them mad. When this happens, they seek refuge on coastal snow patches and higher ground in the mountains.

A quarter of the newborn calves will not live more than a few weeks. Slightly more than half die of birth defects or malnutrition; the number is closely tied to the physical condition of the mother. The rest succumb to golden eagles, grizzly bears and wolves—in that order. Faster, stronger and more wary of the dangers around them, the adults fare much better on the coast. It's on the trek south when the adult caribou are most vulnerable. Along the way, Gwich'in and Inuvialuit hunters in Alaska, the Yukon and the Northwest Territories lie in wait with guns. They and a small number of non-Aboriginal northern residents take as many as five thousand animals, or 4 or 5 percent of the population.

Russell has little doubt that climate change and the growth of the human population in the Arctic is going to put more pressure on this herd, especially if it continues to decline in number due to all of the other factors at play. Most climate models predict that there will be more snow in winter and spring and more heat on the coast in summer that will favor biting flies. How this will affect caribou was made clear on the computer model that Russell and his colleagues created several years ago. The model takes into account population trends based on the size of the herd, cow body weight through the seasons, snow depths and insect harassment. In a worst-case scenario where there are high levels of insect harassment, high snow depths and low body weight among females, the overall caribou numbers fall precipitously.

Needless to say, the Gwich'in of Alaska and the Yukon, who rely on the Porcupine herd more than anyone else, were concerned and called for action when the latest models suggested a further decline in the population. Russell has no idea how many caribou there are now in the Porcupine herd. What troubles him is talk of reviving the strategy of killing wolves in order to allow caribou numbers to recover. Without considering other options, he thinks it's unwise. And like other scientists, environmentalists and many politicians on both sides of the United States–Canada border, Russell is troubled by the energy industry's never-ending effort to convince the U.S. government to allow them to drill for oil and gas in the 1002 Area of the Arctic National Wildlife Refuge in Alaska. This is the last area on the north slope of Alaska completely protected from development.

"Here in the Yukon side of the border, we've done a pretty good job identifying and protecting those areas," he said. "We've created Ivvavik and Vuntut national parks and a special conservation area east of there under the Inuvialuit land claim. But the herd's calving and post-calving range in the Arctic National Wildlife Refuge in Alaska could still be opened up to oil and gas development if the energy industry gets its way. All the research that's been done indicates that this is one of the most important habitats for the herd."

Russell recently retired from the Canadian Wildlife Service but is busier than ever working with the CircumArctic Rangifer Monitoring and Assessment Network (CARMA), a joint effort by university scientists, government scientists and managers, industry, and community organizations to coordinate caribou research and monitoring throughout the world. Russell hopes that research could be used to identify and react to problems before it is too late to do anything about them. As things stand, he pointed out, a lot of information isn't being shared and potentially a great deal of research is being duplicated.

CARMA couldn't have come too soon for wildlife managers in the Northwest Territories, who were doing their best the summer of 2008 to put together a recovery plan for the declining herds within their borders. The news had gone from bad to worse that year when counts suggested that both the reindeer herd in the Mackenzie Delta and the Beverly herd along the Nunavut, Saskatchewan and Manitoba borders were also in serious decline.

With a reindeer population estimated to be as high as ten thousand at the turn of this century, Lloyd Binder, the owner of the herd in the Mackenzie Delta, had plans to increase the number to twelve thousand so that he could harvest two thousand animals each year. Not only would he make money from the sale of meat, he expected to profit from the East Asian demand for reindeer antlers. Many people in the Far East view ground-up antlers as a powerful aphrodisiac.

The new century wasn't kind to Binder's plans. The population has been falling faster than he had hoped it would rise. When he and other reindeer herders went out to count the animals in the winter of 2007–08, they could find only four hundred of the three thousand animals they expected to see.

"They really are cursed," said Binder when I spoke with him in his hometown of Inuvik. "They move around, but we've never had any trouble finding them before. Last winter, the tracks were few and far between and so were the animals."

This isn't the first time that someone has suggested this herd of animals is hexed. Since 1929, when the Canadian government purchased three thousand reindeer from Carl Lomen, the legendary reindeer herder in Alaska, the animals have been nothing but heartbreak for just about everyone who has had responsibility for them. The original goal behind the purchase was to provide Inuvialuit in the western Arctic with a new economic opportunity after the Herschel Island whalers cleared out. A. Erling Porsild, the biologist who was to find a suitable location for the animals, traveled from Nome, Alaska, to the Coppermine River before selecting a site at Kittigazuit, the island now slowly sinking into the Mackenzie Delta.

It would have been a good deal for Lomen had he not agreed to waive payment until the animals were delivered to Canada. What was supposed to be a two-year journey turned out to be a five-year nightmare that ended with only 20 percent of the original animals making it to Kittigazuit. Not that Lomen had been reckless or lazy in getting the animals moved. To make sure the job was done right, he coaxed sixty-year-old Laplander Andy Bahr out of retirement. Bahr then sent for Mikkel Nilsen Pulk and other veteran reindeer herders from Norway, knowing that it could not be done without expert help.

The reindeers' homing instincts were more powerful than the dogteams and whips the herders used to keep the animals in line. Week after week, hundreds would bolt and scatter in different directions. After one particularly brutal July blizzard, it took nearly a full year to round up all of the animals that had ran away. Exasperated by the slow going and overwhelmed by the isolation and harsh conditions, Bahr began to lose his psychological grip. At one point, Pulk and his fellow reindeer herders considered the possibility of a mutiny because they thought he had gone mad.

When the journey was finally completed on March 5, 1935, the Inuvialuit at Kittigazuit greeted Bahr as a hero. And thanks in large part to a couple of years of successful reproduction, he was able to deliver most of the numbers that Lomen had originally promised.

Plans to turn segments of the herd over to local people never succeeded. One of the first groups of animals to be turned over was handed back to the government in 1944 after Charlie Rufus died when his schooner sank in a storm, a disaster that Fred Wolkie told me about the year before in Tuktoyaktuk. Six other family groups also relinquished control over their animals because of grazing conflicts and more attractive jobs opportunities that opened up when the Canadian and American governments set up a series of DEW Line sites across the Arctic in the 1950s. By 1964, it was costing the Canadian government about $80,000 a year more to manage the herd than it was earning from the sale of meat. The government was so desperate to get rid of the animals that it sold them to veteran herder Silas Kangegana for the rock-bottom price of $45,000 in 1974. The price included grazing rights to more than 9,700 square miles (25,000 sq. km) of land.

This proved to be too much even for Kangegana. Some say the heart attack that nearly killed him came from the stress of managing the unruly herd. Whatever the cause, Kangegana was too sick to carry on. He unloaded the animals on William Nasogaluak, an Inuvialuit businessman.

For a time, it looked like Nasogaluak's business acumen would pay off. In short order, he built up the herd to a population of ten thousand. Recognizing the international demand for low-cholesterol reindeer meat and for the blood-engorged reindeer antlers, he turned the animals into moneymakers for the first time. At one point in the 1980s, the herd was generating more than $1 million a year for Nasogaluak's Canadian Reindeer Limited.

This rags-to-riches fairytale turned into yet another nightmare in 1984 when Simon Reisman, the future chief negotiator for the free-trade talks with the United States, settled a land claim with the Inuvialuit on behalf of the federal government. Unwittingly, the settlement placed the most desirable part of the reindeer grazing reserve in the hands of the Inuvialuit, leaving Nasogaluak with ten thousand animals and no place to put them. Nasogaluak, the Inuvialuit and the federal government ended up in a legal tug-of-war that went on for so long, Nasogaluak threatened to slaughter the entire herd at one point. The situation only got worse when the government of the Northwest Territories passed legislation in the 1990s that allowed people to hunt caribou on the grazing reserve. Very quickly, people were shooting Nasogaluak's reindeer on the premise that they looked like caribou.

Lloyd Binder watched this melodrama unfold with quiet fascination. His grandfather was Mikkel Nilsen Pulk, the chief reindeer herder for twenty-five years. Binder was born at Reindeer Station, where reindeer herders were based in the delta. He twice tried to buy the herd in the 1990s but was turned down. When Nasogaluak finally threw up his hands in defeat in 2001, Binder stepped in with a third offer that Nasogaluak could not refuse.

What's become of Binder's reindeer is far from clear. There's speculation that hunters have taken a big chunk out of the population in recent years. Many of the animals may have also dispersed and blended in with the wild caribou of the area, although there's not many wild caribou around either.

Solving Binder's problems was not a priority for wildlife managers in the Northwest Territories. With five of the eight wild herds in decline, they were still busy putting together a recovery plan for those animals. The bad news only got worse when it became evident in the summer of 2008 that the Beverly herd, one of the biggest in the Arctic, was also in a free fall on its range along the Nunavut, Saskatchewan and Manitoba borders. The herd numbered about 276,000 in 1994, the last time a successful census was carried out. Attempts to do a full census in 2007 were foiled by poor weather, but aerial surveys carried out in June 2008 found only 93 females in an area that had produced more than 5,000 in 1994. Worse still were the extremely low numbers of newborns on the calving grounds.

Some Aboriginal hunters saw this coming some time ago when caribou were failing to show up at places where they were almost always seen. Those who pushed for studies to find out where the animals had gone to were opposed by others who objected to satellite tracking collars being put on the some of the animals.

Now that there appears to be a crisis looming, Albert Thorassie, the chair of the Beverly and Qamanirjuaq Caribou Management Board, insists that the research must go forward quickly. He and other board members also want hunters to be vigilant, killing only the animals they need. It's also high time, Thorassie says, to do something about the mining industry that continues to stake claims on caribou calving grounds.

"We've got to get back to the grassroots in teaching our young generation about caribou, and how to protect them," said Thorassie, following the fall meeting. "And the only way is just to take what you need. We're not against mining, but it's important to be careful. Ultimately all courses of action are based on one principle: 'respect, respect, respect' for the caribou."

Anne Gunn was the biologist who headed up the survey that discovered that the Bathurst caribou herd in the north-central Arctic had dropped to 186,000 animals in 2003 from 350,000 in 1996. Calm, soft-spoken and exceedingly polite, she is so much the opposite of Miller that many of their colleagues are astounded that they worked for so long together in the field and produced so many important scientific papers on Peary caribou, muskoxen and their relationship with wolves, weather, the climate and human activity.

What sets Gunn apart from the rest of the crowd is that she was among the first scientists to live in and work in Aboriginal communities while conducting her research. For several years, she was the regional biologist in Kugluktuk, an Inuit community located on the central Arctic coast.

Gunn is now semi-retired, living on Salt Spring Island, but still very much involved in the debate consulting on caribou issues. No one should be surprised that a warmer Arctic may not be good for caribou, she told me the last time we talked, in the fall of 2008. Many of the large mammals of the Arctic—the woolly mammoth, Yukon horses, Alaskan camels, short-faced bears and American lions—all died off over the last eighty-five hundred years when the climate began warming after the last Great Ice Age ended. She believes that the caribou now likely survived that cycle of warming by dispersing and adapting to new habitats. That's why there are distinct sub-species today living in the forests, mountains, mainland tundra and High Arctic.

Now these animals are adapting to another period of warming, which is being intensified by the emission of greenhouse gases, and heading in one direction. What makes the future a potentially grim one for these caribou, says Gunn, reflecting what Miller and Russell both told me, is that the roads, pipelines, cutlines, mines and other human developments are shrinking the size and the quality of the habitat these animals can move in and out from during weather or shifting climatic events. Even worse is that many of these developments encroach on the calving grounds. "We still don't understand the relationship to calving grounds and caribou," said Gunn. "But everything we've learned over the years tells us that it is absolutely critical."

Gunn knows from experience how dangerous it can be to be complacent. In 1980, she conducted a survey of caribou on Prince of Wales, Somerset and Russell islands along the south-central coast of the Arctic. She estimated a relatively healthy population of about six thousand. When she went back fifteen years later to count them again, there were so few animals left that her estimate of one hundred was more hopeful than a reflection of what was actually there.

Try as she, Miller and others did in the years that followed, they had a difficult time figuring out what had happened. There was no evidence that weather was responsible. Nor was there any sign that muskoxen, which were on the rise at the time, had outcompeted caribou for the limited supply of food. In fact, the diet of the muskox is quite different from that of caribou, so that really never was a serious consideration.

It was also unlikely that the caribou moved en masse to another range. The only other explanation was that humans hunted them down to such a low number that wolves and other natural factors prevented a recovery. "It really could have been the sum of a lot of inconsequential things," said Gunn. "The fact is, no one was monitoring the situation. The research funding just wasn't there in a lot of cases to allow us to keep on top of what was going on."

Gunn believes there are lessons to be learned from the complacency that characterizes past management practices. The first is that more needs to be done to identify and understand the causes of declines so that conservation measures can be taken before it's too late. The Inuit and First Nations people must also be directly involved in survey efforts and other studies so that they can buy into a strategy that might require them hunting less animals. Not doing this in the past, she notes, continues to haunt biologists and wildlife managers all across the North.

"One thing that is certain is that time is not on the caribou's side," she said. "We cannot afford to dither, given the rate of changes we are unleashing across the Arctic regions."

Most everyone agrees that it's going to be extremely difficult to get people to think differently before it's too late. This became apparent in 2007 when the Mackenzie Valley Environmental Impact Review Board, one of the regulatory agencies in the Northwest Territories, rejected Ur-Energy's proposal to explore for uranium at Screech Lake in the upper Thelon River region northeast of Great Slave Lake.

This is the same part of the world that Mike Ilnik talked about when we crossed paths in Rankin Inlet in the spring of 2007. The Dene believe it is the "place where God began." Because large numbers of caribou come here each year to have their young, they wanted the exploration stopped before a mine got started.

Not everyone in the North was happy with the review board's unprecedented decision to reject Ur-Energy's proposal on cultural and spiritual grounds. The mining industry, in fact, was so incensed that three of its umbrella organizations—the Mining Association of Canada, the Northwest Territories and Nunavut Chamber of Mines and the Prospectors and Developers Association of Canada—appealed directly to Canadian prime minister Stephen Harper to reject the decision, adding that failure to do so would "hasten the flight of mineral exploration dollars from the N.W.T. to other, more receptive, parts of the country."

When I met Gabrielle Mackenzie-Scott, the review board chair, in early 2008 to talk about the decision, she didn't look the least bit like the grandmother she had become in recent years, nor the mother of eight children she has been for a good part of her adult life. Nor was there anything in her past to suggest that this tiny woman from a small Dene town 50 miles (80 km) northwest of Yellowknife could intimidate or frighten powerful resource industry interests in the Arctic.

"I grew up in the small community of Rae and lived there until I went to school in Yellowknife," she said. "So I know what's it's like for people in these small places to have to deal with all the pressures that are now coming with mining and oil and gas development. It can be overwhelming, especially when you're trying to protect something that has as much cultural or spiritual value as it has economic value."

The prospect that people have a connection to the land beyond the physical landscape is something that Mackenzie-Scott sees as relevant. The problem, she says, is that most small communities in the Northwest Territories simply don't have the resources to make that case. Not only are they struggling to deal with the proposed $16-billion Mackenzie Valley gas project, they are also trying to cope with diamond drilling ventures, uranium developments and oil and gas explorations that are literally taking place in every corner of the North.

"Overworked, underfunded and understaffed" is how she described the plight of many communities she dealt with during her term as chair. Despite being criticized for the Ur-Energy decision, Mackenzie-Scott was unapologetic. "To the people who hunt and trap there, it's 'The Place Where God Began,'" she said. "We heard from everyone, a fourteen-year-old boy to a ninety-seven-year-old elder who talked about the spiritual importance of this land. Is it wrong to protect a place that is important in this way? Doesn't everyone have a place they consider to be sacred and worth preserving?"

During the time we talked, Mackenzie-Scott said she wouldn't rule out the possibility that she and the board would make similar decisions based on the cultural and spiritual factors. That option, however, was taken away from her when, a month later, the Canadian government rejected the board's recommendation that she be reappointed to a second term.

chapter nine

RICH SQUIRREL, POOR SQUIRREL

– Mile 1004 Alaska Highway, Yukon –

This is the belt of dwindling trees, the last or northernmost zone of forest, and the spruce trees showed everywhere that they were living a life-long battle, growing and seeding, but dwarfed by frost and hardships. But sweet are the uses of adversity and the stunted spruces were beautified, not uglified, by their troubles.

—Ernest Thompson Seton, *The Arctic Prairies*, 1911

ALONG A LONELY STRETCH OF ALASKA HIGHWAY in the southwest corner of the Yukon, an impossibly rough sideroad stops dead a few hundred feet into the forest. Among the spruce and aspen trees are a mess of ramshackle buildings and trailers. The only sign of life on this day in late May is a wisp of smoke spiraling from a chimney and a red squirrel, a collar around its neck, running across the roof.

Anyone who stumbles onto this place might think he'd discovered hillbillies or squatters. There's no telling what he'd think if he were brave enough to carry on and step over the electric fence and read the sign that says, "Remember Poop!"

Quinn Fletcher would only add to the intrigue should he politely answer the door of one of the rickety buildings as he did on the evening of my arrival. Sporting a fleece hat that sat higher on his head than likely intended, he welcomed me before excusing himself so he could process some squirrels' blood that he had packed away in a thermal lunch bucket tucked under his arm.

And so a typical scene unfolds at "Squirrel Camp." University of Alberta biologist Stan Boutin and other scientists from Canada and the United States have been operating this remote research station, located along the border of Kluane National Park, for more than twenty years. Here in a world saturated with the solitary chatterboxes of the boreal forests of the sub-Arctic, there may be as many as fifteen scientists and students running around at one time, drawing blood, collecting poop, raiding nests, locating middens (food caches), counting spruce cones and pouring peanut butter into plastic buckets that they hang high in the trees. With the peanut butter as bait, Boutin and his colleagues have trapped, blood-sampled and followed the fate of nearly six thousand squirrels since 1987, often with the help of ear tags and tiny radio transmitters that they wrap around the animal's neck.

The scientists at Squirrel Camp have gotten to know this population well. They can say who is stressed, who is shy or adventurous, who's a bully and who's the brother or sister, daughter or son of so-and-so from each new generation. They know what becomes of a family of squirrels that's given handouts and what it is that squirrels are communicating when they rattle, squeak and chatter, as they so often do. They're not just doing this to understand the nature of squirrels. They're doing it to answer a number of fundamental biological questions about stress, metabolism, personality and genetics. Boutin and his peers want to know how these factors play out in a world that is constantly being transformed by land fragmentation, forest fires, disease, natural variability and climate change.

When Fletcher greeted me at the door, Stan Boutin was just about to eat supper with a group of students who had come out of the field. The squirrel with the collar around its neck was sitting on a branch, watching them through the window, as if expecting a handout. An orange tag was on its ear and a wire antenna loop stuck up from a tiny transmitter attached to its collar. For a second, the squirrel looked like a mechanical toy.

The facilities at Squirrel Camp are even more rustic than the bigger camp that pilot Andy Williams manages down the road at Kluane Lake, where I sat for several days the year before, and once again this year trying unsuccessfully to fly into David Hik's pika research site in the icefields and the Ruby Range. Like at Kluane, this camp has the warmth and smell of home cooking and the pleasant atmosphere that comes from people who seem to be engaged in their research and having fun at the same time.

I didn't need to introduce myself to Boutin. He and I had met several times before. In his fifties, he still looked as fit as he was when I had last seen him. He attributed his good health to the necessity of having to keep up with two high-performance soccer daughters, the good genes he inherited from his parents and an active schedule in and out of the research field.

"We tried working out of the Kluane Research Station early on, but then decided to move out here to save on the time and gas," said Boutin as he passed me a beer from a cooler that, I was relieved to see, didn't contain squirrel droppings or blood. "We used tents for a while, but that got to be a little uncomfortable when we started pushing the work into the winter season. So I figured out a way where we could put these buildings up. It's not luxurious, but it's much appreciated at the end of a long, cold or wet day in the field."

In the competitive world of wildlife research, Squirrel Camp is a bit of an anomaly. Unlike caribou and polar bears, which have enormous economic and cultural value, there's no pressing need to study these animals. The red squirrel is not endangered, threatened or even at risk. Nor is it a keystone species on which predators rely. The squirrel is just too fast and wily to be a dependable lunch for wolves, marten, goshawks, owls and other predators that occasionally catch it. It is, however, a good backup for lynx when the snowshoe hare population goes from boom to bust. The downside for lynx is that it takes six squirrels to get the same energy that a meal of hare provides. The upside for squirrels, if it can be called that, is that this temporary shift in feeding strategy doesn't appear to have any serious impact on its population.

The money for Squirrel Camp keeps pouring in largely because of the high-profile scientists associated with the project. In addition to Boutin, who holds a prestigious Canada Research chair for studies he is doing farther south, people such as Rudy Boonstra with the Centre for Neurobiology Stress at the University of Toronto, wildlife energetics specialist Murray Humphries of McGill University in Montreal and evolutionary biologist Andrew McAdam of the University of Guelph all bring impressive talent and reputations to the table. It doesn't hurt that scientists such as Dominique Berteaux of the Université du Québec à Rimouski and Charley Krebs, the ecologist who literally wrote the book on rodent and snowshoe hare ecology, have been associated with the group over the years.

"I'm not sure there's anything quite like our camp in the North," said Boutin, tucking into a vegetarian plate. "Part of it has to do with timing and the talent we have. I doubt that we'd get very far today if we were just starting off and going to funding agencies asking for money to sustain a long-term project like this. Things are getting a little better, but there's such a big emphasis on producing immediate results that are of economic value that most biologists stay away from research that will last for more than five years. It's unfortunate because it's virtually impossible to get a handle on how climate change, land fragmentation and other environmental factors are affecting wildlife species without long-term studies like ours to go on. Most of what we've learned here has evolved from research that got its start more than twenty years ago."

The fact that red squirrels are thriving in a rapidly warming world is significant considering how grim the future looks for so many other birds and animals. The 2007 report from the Intergovernmental Panel on Climate Change (IPCC), which was written and edited by more than twenty-five hundred scientists and reviewers from one hundred and thirty countries, projected that up to 30 percent of all animal and plant species could be wiped out by a global temperature rise of about 2.5 to 4.5 degrees Fahrenheit (1.5 to 2.5 degrees C), which is expected to occur well before the end of this century. Andrew Weaver, the lead author of the first IPCC report, likens the path on which the world is heading to a "highway to extinction."

In the Arctic, where temperatures in most places are rising faster than they are anywhere else, the outlook is even grimmer, as climatologist Gordon McBean succinctly pointed out in a landmark report that demonstrated how volatile and unstable the Arctic climate can be without the added input of greenhouse gas emissions.[1] The reasons why some animals are disappearing or dwindling to dangerously low numbers were made obvious to me on my previous trips to the Arctic. Without a platform of ice from which they can hunt seals, polar bears at southern and midpolar latitudes are all but doomed. Caribou will find it increasingly difficult to migrate through deep snow that warmer spring weather is likely to bring. They will also find it hard to find lichen, a primary food source, in forests that will increasingly be destroyed by bigger and hotter fires.

More often than not, the impacts of warming on wildlife are and will continue to be less obvious. Biologists Gordon Court and Alastair Franke speculated that peregrine falcons, once endangered by PCBs and dioxins in the environment, would do better on their Arctic nesting grounds because a warmer climate promised to bring an abundance of prey. They were right on one count: the prey base in their study area on the western coast of Hudson Bay did increase. But the wet weather that accompanied the warmer temperatures produced unexpected results. The heavy rain soaked the downy feathers of peregrine chicks to the point where the chicks eventually succumbed to hypothermia. In the summer of 2005, virtually every nest that Court and Franke were monitoring failed to produce healthy birds by the end of the season.

In other cases, the cause and effect of climate change can be downright confounding. Biologists are certain that climatic factors are playing a role in the decline of barren ground caribou. Yet no one has figured out

1. *Arctic Climate Impact Assessment* (2005) was prepared by an international team of more than three hundred scientists, experts and knowledgeable members of indigenous communities.

why it is that numbers in the George River herd in northern Quebec have plunged from 800,000 in 1993 to 440,000 in 2001 when the neighboring Leaf River herd has almost doubled during that time. A similar situation is occurring in Alaska, where the western Arctic herd appears to be stable, but the Porcupine herd that Don Russell has been studying has been steadily declining.

Then there are those birds and animals in the Arctic that are proliferating in a warmer environment. In recent years, the barren ground grizzly, an animal that is about two-thirds the size of its cousin in the Rocky Mountains, has been expanding its Arctic range east toward Hudson Bay, where it was once a rare sight. Some of these animals have also been moving north onto the Arctic Islands and into the icy kingdom of the polar bear, where there's evidence that the two animals have fought, mated and produced at least one hybrid. Scientists are still amazed that the grizzly bear spotted by Earth scientist John England on Melville Island in the High Arctic in the summer of 2003 was still there when Charles Francis surveyed the island four years later. No grizzly has ever been seen so far north of its traditional range.

No one knows for certain whether climate change is behind any of this, but Andrew Derocher, the scientist who has been working on polar bears with Ian Stirling in the Beaufort Sea, has a theory. He and his students are in the midst of doing a long-term study of coastal grizzlies in the Northwest Territories and the Yukon. As he pointed out to me when I spent some time with him in the field, with spring coming earlier and winter coming later, barren ground grizzlies do not have to hibernate as long as they once did. This gives them more time to put on the fat they need to successfully reproduce and get through a long winter.

The proliferation of snow geese, on the other hand, has just as much to do with what's happening on the wintering grounds in the south as it does with what's going on in the Arctic. Rice and corn crops that have taken over wetland areas in the southern United States are providing geese with a rich and plentiful supply of food. As a result, more geese are surviving the winter. Those that migrate north are healthier and better able to successfully reproduce.

The explosion of snow geese may sound like a good-news story, but it is rapidly becoming an environmental disaster. So many geese are now nesting in the Arctic that vast stretches of lowland tundra are being desertified because geese pull up the shoots of the plants by the roots instead of snipping them at the top. These denuded areas quickly become alkaline and toxic to many Arctic plants that attempt to re-establish themselves in the area. What's more, those denuded areas are no longer useful for caribou, muskoxen and other animals.

Boutin is not surprised that there are many shades of gray, a lot of black and white and a number of unknowns in the climate change scenario in the Arctic. "I have always thought that the winners will be the usual suspects: organisms with short generation times that have the potential to adapt quickly, those that occupy habitats that will be favored or at least won't disappear, and those lucky enough to not fall victim to new disease or parasites that come with the change," he told me that night.

"But it's not all that simple, as our research on squirrels demonstrates. We're not here specifically to answer all the questions that climate change poses. But I think there's something to be learned from these and other animals that are adapting so well. The more we know about an animal, the better we are at managing them."

Red squirrels are one of eight tree squirrel species found in North America. In Canada and the United States, they are commonly found most anywhere conifers grow. The seed of the spruce cone is the favored fruit in the animal's diet. Being no more than 13 inches (33 cm) long, the red squirrel is hardly a formidable presence in the forest. It is, however, as Boutin the bike racer likes to say, "the Tour de France athlete" of the animal kingdom.

The red squirrel is a lean and seemingly tireless machine that can crank up its metabolic basal rate to levels that only Lance Armstrong and a small number of world-class athletes are capable of doing for a sustained period of time. Basal rate is the minimal amount of energy an animal expends while sleeping or resting. Exercise physiologists have shown that the energy expenditure of athletes such as Lance Armstrong is about four times this average basal metabolic rate. It's the same for squirrels during the mating season: all hell breaks loose and male squirrels run around frantically, trying to mate with the relatively few females that are sexually receptive in early spring.

For student Ryan Taylor, who must snowshoe after the frenzied males to keep track of this mating ritual, it can be brutal work. Because the female is sexually receptive for just one day of the year, males go all out to make sure they don't miss their chance. They can cover a lot of ground. It's the one time of year they switch off the early warning signal that tells them whether a predator is nearby.

"When there's snow on the ground, they look like otters swimming," said Taylor, who was listening in on my conversation with Boutin. "They are relentless in the chase, and it's very, very difficult to keep track of individuals because more often than not, there's more than one male chasing a female. And when they copulate, they'll do it in trees, under the snow, in the hollows of a stump, almost anywhere."

The red squirrel is also set apart from most other small animals in the boreal world because it does not hibernate. Nor does it put on copious amounts of fat, which is remarkable for such an energetic creature living in a world where temperatures can dip from -22°Fahrenheit to -40°Fahrenheit (-30°C to -40°C) for weeks at a time. Normally, bigger is better when it's cold for so long.

Red squirrels also possess an uncanny ability to anticipate when a spruce tree will randomly produce an overabundance of spruce cones. This food surplus enables a squirrel to produce two litters in a summer instead of just one. The red squirrel's ability to sense when a tree will be overproductive is remarkable because most trees have evolved in a way that makes the timing of this "swamp and starve" strategy extremely random and unpredictable. The extra cones ensure that in some years there will be enough seeds remaining at the end of the season to take root the following year.

"We first got a hint of this back in the early 1990s when we were caught off-guard by the huge numbers of pups that the squirrels were producing in our study area," said Boutin. "Students were being run ragged that summer trying to catch and tag all of them. We couldn't figure out what was going on and thought it would never end."

Boutin doesn't yet know how the red squirrel anticipates when there will be a bountiful harvest. The female pretty much has to make the decision about how many young she's going to produce in February or March. The spruce cones don't mature until late August or September.

As we headed outside for a walk toward the forest's edge that evening, Boutin pointed up into the trees. "It takes about eighteen months for a tree to produce a seed cone, starting from small buds at the ends of the branches that will grow into either branches or cones. We know that the squirrels feed on these buds. It's possible that they're counting them early in the season to see if they will have the food necessary to produce the extra litter. It's also possible that there's something in the bud that tells the squirrel which is going to produce a branch and which is going to produce seeds."

Boutin did not intend to study squirrels. When he was a graduate student at the University of British Columbia, he focused on the population dynamics of snowshoe hares and its predators. The relationship between hares and lynx and other predators had been a source of scientific and economic interest in North America since 1831, when a Hudson's Bay Company manager wrote to England describing how a shortage of hares seemed to be related to a paucity of fur-bearing animals such as lynx and wolverine.

That observation was borne out by two hundred years of fur trade records that demonstrated how these cyclical fluctuations occurred without fail approximately every ten years. Boutin's original game plan was to find a way of better understanding the mechanisms responsible for the booms that could result in two hundred to three hundred hares per square kilometer (0.3 square miles) and the busts that saw those numbers drop to seven or ten.

Capturing adult hares wasn't so much a problem, but finding youngsters was. So much time elapsed from the moment of birth to the point when Boutin was able to find and trap the young animals that he was left with a big gap in his understanding of the animal's life cycle. This, of course, was bad news for a young scientist trying to make his mark in the academic world and secure enough funding to keep his field studies going. In fairly short order, Boutin started looking around for another, more cooperative animal.

Boutin remembers the day he came up with the idea of shifting his study to squirrels. "It was just one of those things," he said. "I was out there in the forest and this squirrel was not at all happy about me being in its territory. I'm used to this because they're always chattering. Usually, I ignore it. Somehow, this guy got to me and that's when I realized that maybe it might be the perfect experimental animal. The more I thought about it, the more it made sense. They are long-lived and there are lots of them. They live in a small space and you can follow them from birth to death. You can also manipulate their world in many different ways without doing them serious harm or driving them off."

The following morning, Quinn Fletcher took me to one of several nests he and other students were monitoring. This one was down the road about a half-mile (1 km) off the Alaska Highway alongside an old cutline. To help him find the exact spot, he carried an antenna above his head to home in on the transmitter secured around the neck of the animal earlier that spring. The transmitter, which weighs just 0.15 ounces (4 g), transmits a pulse at a specific frequency every five seconds. The antenna amplifies the sound of the pulse as Quinn draws near. The sound gets softer whenever he veers away.

In the event that we bumped into a bear, Fletcher and I both had a can of pepper spray ready. Along the way we passed one of the sites where students have been feeding the squirrels peanut butter. This ritual begins in September and continues through to April/May when it is still cold. To soften the peanut butter, the students heat the cook shack up to 95°Fahrenheit (35°C). They then pour the peanut butter into 2-pound (1 kg) yogurt containers, trying not to spill and mess up the floor. This happens more often than they'd care to admit. A select number of squirrels get one of these containers every two weeks in winter and spring. The animals' data is then tracked and compared to squirrels not provided with the extra rations. The aim is to see whether this additional food supply affects survival and the number of pups a female will produce.

Typically, a female red squirrel will produce four pups in nests she constructs out of grass and witch's broom. Invariably, she will hide these nests in a spruce or poplar tree or farther down in the cavity of a trunk. The young, blind at birth, stay in the nest for about forty-two days, or until they are strong enough to venture out on their own.

Females are very good at disguising their nest sites. Even when Fletcher figured out which tree housed the squirrel's nest, it was almost impossible to tell from ground level where the nest was. "You see that cluster of witch's broom?" said Fletcher. "That's where she's sitting. When I climb up, she'll race to the top of the tree and there will be a lot of chattering. But you can be sure that she'll have her eye on me every second."

And so she did. After gingerly climbing up the tree, Fletcher carefully reached in and, one by one, slipped all four pups into a thermal bag he had draped around his shoulder. On the ground, he took one the size of a mouse out and let it squirm and squeak until it had no energy left. Then, with a surgeon's precision, he used a crimper to clip a tag onto the pup's ear. The pup was too tuckered by this point to wake up and complain.

"It will be interesting to see what she [the mother] does when I put them back into the nest," he said. "Sometimes they pick up and go off to some other place where there will be another nest site. But she has to be careful to make sure that there's no other squirrel using that territory."

It didn't take long for the female squirrel to act. Seconds after Fletcher was on the ground, she was climbing down the back of the tree with a pup in her mouth. Ten minutes later, she returned for another. Within the hour, there was no sight of her or her new nest.

Big as the forest is, finding real estate is the only serious problem a young squirrel faces in the wild. When it comes time to leave the nest, new homes and middens are hard to come by. More than one in five young squirrels end up homeless, falling victim to malnourishment, sickness or a violent death. A female squirrel will go to great lengths to protect her pups from such fates. Remarkably, not only do mother squirrels bequeath middens to some of their offspring, they also try to acquire additional middens long before the young are even conceived.

Boutin calls this phenomenon, never seen in another animal, the "silver spoon" effect. He likens it to parents putting away money for their child's university education even before they plan to be a parent. The females acquire these food caches. When it's time for the kids to move out, they turn the supplies over to the young squirrels. Nothing the scientists have observed so far suggests the female gets anything in return.

Boutin and his colleagues discovered that the choices made by a squirrel's mother played a significant factor in its fate. Pups born under favorable conditions, be they bequeathed middens, warm weather and years when spruce cones are in good supply, appear to fare better in the long term than those born in conditions where the weather is severe and the food supply scarce. Boutin and Andrew McAdam were able to manipulate the environmental conditions to determine whether it mattered if the pup was raised by a "super mom" or a lazy mother. Environmental factors, they discovered, are important. But by doing things such as taking one pup and placing it into another nest, the scientists discovered that the choices the mother made were also important.

In the world of biology, scientists have a tendency to pigeonhole various species into behavioral roles that range from the aggressive to the passive. Traditionally, the assumption is that the aggressive animals are the most successful in securing a mate, finding food and fending off predators or rivals. Although this in many cases is true, growing evidence suggests that it's not always the case, especially in a natural world being transformed by human activities. In the Rocky Mountain national parks, it's becoming increasingly apparent to some biologists that the more passive grizzly bears that stay away from people and traffic are the ones that do best in the long run because they are less likely to get shot, hit by cars or euthanized when they demonstrate any aggressiveness toward humans.

This notion of personality, and how that plays out in the life cycle of an individual animal, is another topic Boutin is exploring. By observation, the Squirrel Camp scientists know that squirrels are like most other animals and even humans: some are highly aggressive and others are more laid-back and passive. To get a better handle on this spectrum of squirrel personalities, Boutin got Adi Boon, one of his graduate students, to measure these personality traits at Squirrel Camp.

Each day Boon would trap and release a squirrel into an arena—a big box with holes in the floor, a reflective mirror at one end and a video camera at the other that records the animal's response when it sees and does not recognize its own image. Invariably, the passive ones would hang back or ignore the image. The aggressive ones would attack.

Boon and others would then track these individuals to see how they fared in the wild. Although there are advantages to being aggressive, Boon has also found that it has its drawbacks. In bumper-crop years, the pups of aggressive squirrels, which were able to obtain greater amounts of food than their passive peers, grew faster because the extra food helped their mother to nurse and provide them with spruce cone seeds when they were older. In lean years, the pups of passive squirrels fared better. Being aggressive doesn't necessarily mean the A-type male always gets the girl either, as one might expect. Studies at Squirrel Camp suggest that the thoughtful, passive fellows are just as successful.

Boutin and his colleagues are also focusing on the role that stress plays on the health and survival of the animal. This is particularly important in an environment affected by a climate change scenario, where one or two new stressors introduced in the life cycle of an animal make all the difference to that animal's survival. Canadian Wildlife Service scientist Tony Gaston saw how just seemingly small changes introduced by climate change had serious repercussions for thick-billed murres he was studying on Coats Island in northern Hudson Bay. In that area, thinning ice caused by warming temperatures favored fish-like capelin and sundlance over arctic cod and sculpin, which traditionally dominated in those icy waters. Gaston saw this shift manifest itself in the diets of thick-billed murres. Although the murres appear to be adapting, they have to expend more energy catching these fish. In addition to altering their hunting strategies, murres are having to cope with mosquitoes now hatching earlier in the season. Unfortunately, the birds have not been as successful in coping with this change. Some nesting birds are literally getting blood sucked out of them to the point where they perish.

Red squirrels are likely going to have a difficult time adapting in some parts of the sub-Arctic, especially if climate change continues to bring lightning and warm, dry weather that has been triggering large, hot fires in the Yukon and Alaska in recent years. The potential is there for these fires to burn right into the ground, where they can destroy the spruce seeds required to rejuvenate the forests.

Yet, like long-distance bike racers, red squirrels seem to be built to cope with stress. This is especially evident during the breeding season, when a male squirrel must crank up an aggressive attitude to stay competitive in the marathon race for a mate while defending its territory. They maintain this brutal pace by pumping out testosterone, a predominately male hormone that not only stimulates sexual activity, but also builds muscle mass and bone density. Although testosterone gives a squirrel the drive to go on an epic chase that can last eight hours on a cold winter day, the hormone is also potentially lethal because it burns energy at a rapid rate and weakens the immune system—not a good thing for such a lean animal that lives in a very cold environment.

Given the fact that squirrels are just as aggressive in defending their territory in the non-breeding season, Rudy Boonstra, the University of Toronto scientist who has devoted his entire career trying to understand how stress plays out in animal life cycles, wondered a few years back how it was that squirrels do not suffer a downside, especially when they have so little fat to burn. By analyzing blood samples provided by Boutin's team, he found that in the non-breeding season, squirrels make the switch from testosterone to DHEA, a natural steroid produced by the adrenal gland. It's what some non-breeding songbirds circulate during periods of aggression. The steroid is also related to testosterone-like compounds such as stanizol, which Ben Johnson used to win the gold medal in the Olympics at Seoul. DHEA is now banned by the World Anti-Doping Agency.

Circulating DHEA rather than testosterone would certainly be an advantage for the red squirrel, especially in winter when the food supply is limited, the temperatures are cold and the squirrel has to protect its food store. DHEA allows the squirrel to maintain its aggressiveness without burning excessive amounts of energy and compromising its immune system.

Although DHEA may be one reason why squirrels are so well adapted to living in a boreal environment, it's genetics that has given them this and other tools required to deal with the catastrophic changes taking place in its world. Over the past three decades, climate change has triggered a number of events in the Kluane area, including spruce beetle outbreaks, earlier budding of trees and predator-prey cycles that have seriously challenged other boreal birds and animals. Southern animals such as the coyote are also moving into northern territory and competing for a limited prey base. Unlike thick-billed murres, which are struggling with the changes to some of their nesting territory, red squirrels seem to be adapting extremely well. In response to the earlier arrival of spring, females are producing litters eighteen days earlier than their grandparents did.

Because red squirrels are such a good experimental animal model, Boutin and his colleagues are now manipulating their world in other ways to see if a similar evolutionary response occurs. This is where the peanut butter comes in. Andrew McAdam and his students are feeding some squirrels more peanut butter than the creatures could ever hope for. In the coming years, the scientists will compare how these so-called rich squirrels are faring relative to the poor squirrels that must cope through lean years.

Even with more than twenty years of research, Boutin doesn't see Squirrel Camp closing down any time soon. The possibilities for new ideas, he says, are almost limitless. "It's a rare thing to be able to track evolution in a wild population of animals," says Boutin. "We're pretty unique with our set-up. So far, climate change has been a good-news story for these squirrels. But the change is also happening very fast and the warming this time is moving in one direction, where in the past, a period of cooling tends to follow. It remains to be seen whether these animals can keep up with the pace of change in the future."

chapter ten

THE COLDEST WAR

– Canadian Forces Maritime Warfare Centre, Halifax –

If you wish to know what men seek in this land, or why men journey thither in so great danger to their lives, then it is the threefold nature of man which draws him thither. One part of him is emulation and desire of fame. . . . Another part is the desire of knowledge . . . the third part is the desire of gain.

—From *The King's Mirror*, by an unknown Norseman, circa 1250

In the summer of 2007, the Russian government sent two icebreakers to the North Pole on a mission that was supposed to be mainly scientific. Only when the ships got to the top of the world did the geopolitical purpose of the venture become apparent to the rest of the world. Veteran Arctic explorer Artur Chilingarov, a member of Russia's lower house of parliament, descended 14,000 feet (4,300 m) in a deep-sea submersible and deposited a Russian flag, cast in rust-free titanium, on the sea floor.

Clearly, the entire event was choreographed and filmed to announce to the world, and to the Russian people, that the seabed under the Pole, the 1,200-mile-long (1,900 km long) Lomonosov Ridge, was an extension of Russia's continental shelf. Expedition members were treated like heroes when they came home. "We were there first and we can claim the entire Arctic, but if our neighbors want some part of it, then maybe we can negotiate with them," said Vladimir Zhirinovsky, the populist leader of Russia's ultra-nationalist Liberal Democratic Party.

Despite how audacious Zhirinovsky's comments seemed, given the rules and legal regime in place for resolving boundary issues in the Arctic, they reflected in some measure what the Russians were thinking. A few days after the flag planting, strategic bombers were dispatched over the Arctic Ocean for the first time since the end of the Cold War. "The division of the Arctic," the Russian daily newspaper *Rossiiskaya Gazeta* declared sometime later, "is the start of a new redistribution of the world."

Canada was the first—but not the only—country to protest. "This isn't the fifteenth century," said Peter MacKay, the country's foreign affairs minister. "You can't go around the world and just plant flags and say, 'We're claiming this territory.'" The United States was a little slower off the mark. State Department officials appeared to be taken off-guard by the theatrical nature of the move. The United States wasn't alone in wondering what to make of it all. Most countries with interests in the Arctic seemed puzzled by what the Russians were up to.

The following summer, I was sitting in the conference room at the Canadian Forces Maritime Warfare Centre listening to naval and security experts discussing Russia's polar ambitions and strategizing over other maritime security issues. On one side of me was Rob Huebert, the associate director of the Centre for Military and Strategic Studies at the University of Calgary. On the other side was Martin Murphy, a research fellow at the Corbett Centre for Maritime Policy Studies at King's College in London.

Danford Middlemiss, the Dalhousie University director of the Centre for Foreign Policy Studies, had already reminded everyone that we should be "humbled" by the shortcomings of past predictions when it comes to securing boundaries. Who, he asked, saw the end of the U.S.S.R. coming? The need now, he said, was to consider the so-called "invariables" that are bound to challenge national security as "drastic and revolutionary changes" unfold in the Arctic and the rest of the world.

Canadian senator Colin Kenny, chair of the Standing Senate Committee on National Defence and Security, likened Canada's Arctic policy record to that of the United States and the United Kingdom during the 1930s. "Our country is asleep," he said after describing how climate change was opening up Arctic waters to foreign ships. "We underestimate the threats around us."

Ultimately, it was left to Jack Granatstein, the venerable Canadian military historian, to put the issue into perspective. "We are not going to go to war over the Arctic in the future," he reassured everyone at the beginning of his presentation.

"But we might."

Huebert might have been forgiven for gloating a bit that day. For nearly a decade, he and Colonel Pierre Leblanc, the former commander of the Canadian Forces Northern Area, were virtually alone on center stage warning both the Canadian and U.S. governments that the rapidly melting ice in the Arctic was as much a threat to sovereignty and security of the continent as it was a huge economic opportunity for oil, gas and mining interests.

Each time a crisis arises in the Arctic, Huebert told me during a break in the discussions, the government gives the public the impression it's going to act expeditiously to assert sovereignty and security. Yet whenever it comes time to committing the resources to actually doing something meaningful about it, the government backs off in the hopes that the issue will go away. The Canadian government has been able to get away with that in the past, said Huebert, but he's certain that they're not going to get away with it in the future.

The Russians planting a flag at the North Pole wasn't the first time the issue of sovereignty and security in the Arctic has captured the attention of North Americans. Security and control in the Arctic dates back one hundred years when Joseph-Elzéar Bernier planted a flag on Melville Island, claiming the entire Arctic Archipelago for Canada. The issue resurfaced for a while in the 1940s and 1950s, when the United States built the Alaska Highway from Dawson Creek in northern British Columbia to Fairbanks, Alaska, then the Canol oil pipeline from Norman Wells in the Northwest Territories to the Yukon and finally a string of Distant Early Warning (DEW) Line sites that extended from Alaska to Baffin Island.

For a time, the Canadian government was content to see the Americans footing the bill for these defense- and security-related projects, but the growing number of U.S. flags on Canadian soil eventually became an embarrassment. The Canadian government responded to public concern by fortifying RCMP presence in the North, sailing a few Coast Guard ships through Arctic waters and, in a move that future government leaders would come to regret, relocating Inuit from northern Quebec to two uninhabited islands in the High Arctic. When public pressure dissipated, both politicians and bureaucrats retreated to Ottawa and forgot about it.

The United States has never recognized Canada's sovereignty over the Northwest Passage or the offshore boundary line that Canada has drawn between Alaska and the Yukon in the western Arctic. In 1969 and again in 1970, the U.S. oil tanker *Manhattan* challenged Canada's sovereignty claims by sailing through the Northwest Passage without seeking Canadian permission. When the *Polar Sea* did it in 1985, it triggered a national debate that culminated with the Council of Canadians, a group of prominent citizens committed to economic and cultural sovereignty, dropping a Canadian flag on the American ship as it sailed through the Arctic.

The ensuing controversy led to high-level diplomatic discussions between Canada and the United States. Joe Clark, the country's foreign affairs minister, reassured the public that Canada's sovereignty over the Arctic is "indivisible. It embraces land, sea and ice," he said before holding diplomatic meetings with his U.S. counterpart. "The Arctic is not only a part of Canada," he said in the same speech, "it is a part of Canadian greatness."

The most Clark could extract from the United States in 1988 was a promise that the Americans would at least serve notice when they intended to attempt a transit of the Northwest Passage, waters they and other nations consider to be an international strait.

That hasn't always happened. Several American submarines have since made the journey unannounced, sometimes playing cat and mouse with Soviet and now Russian subs that were doing the same thing. The French and British have been at it as well. At least twice in the past, the Canadian government tried to allay public concern about these territorial incursions by proposing to build nuclear-powered icebreakers. Both times it backed down because of the high costs and because public interest in the issue waned.

When Canadian prime minister Stephen Harper announced in August 2007 that Canada was going to build a new Polar Class icebreaker, five to eight Arctic naval patrol vessels and refurbish an old mine port at Nanisivik along the Northwest Passage, military and security experts were encouraged. But Huebert wasn't entirely convinced that the government would make good on these promises. If history tells us anything, he told me, it tells us that these things tend to slide off the priority list when issues such as the economy come into play. This, he said, is a very good start. But until contracts are signed, he was reserving judgment.

For Colonel Leblanc, now retired but still intimately involved in the sovereignty and security debate, Harper's announcement was long overdue. Leblanc was commander of the northern forces from 1995 to 2000. Throughout that time, he witnessed a number of things that made him worry about Canada's ability to deal with security threats.

One of the first and most brazen security breaches occurred in the fall of 1998 when a Russian IL-76 flew over the North Pole en route to the port of Churchill. Mike Lawson, the man on airport duty, remembers it well. The IL-76 is an unforgettably large cargo plane, even bigger than the enormous C-130 Hercules used by the Canadian military.

"We don't get big Russian planes like that in Churchill," he told me when I talked to him shortly before this conference. "In fact, in the eighteen years I've been here, I've seen only one other like it."

It was a dirty night weather-wise. Strong winds were blowing snow on the runway, so visibility was marginal at best. That's not unusual for November in this part of the world. What was unusual was the pilot switching off his landing lights the moment he hit the tarmac. That made

Lawson wonder how many times the pilot had landed in Afghanistan, where the practice was a common way of preventing landed planes from becoming easy targets.

The Russian crew was not an "overfriendly" bunch when they disembarked, but their somber mood had changed dramatically the next morning. Local helicopter pilot Steve Miller crossed paths with them at Gypsies, a popular restaurant in town. Miller had spent a good part of his career in the North piloting polar bear scientists such as Ian Stirling and helping them catch and process the animals on the ice. I had also spent some time with him and biologist Peter Clarkson in the field and gotten to know him reasonably well.

Miller remembers the Russians, mainly because it was 10 a.m. and they were drinking beer. When the Russians discovered that he was a pilot, they tried to get him to drink with them, claiming that a real pilot would be drinking at that time of day as well. Miller declined.

The Russians didn't stay long. Shortly after a Bell 206 helicopter landed at Churchill, the Russian crew drove back to the airport, dropped the plane's cargo doors, loaded up the helicopter and took off.

"Just like that," said Lawson. "No one was there to ask questions or inspect documents. It makes you wonder who's guarding our back door."

What Lawson did not know then is that Canadian intelligence officials were monitoring the flight of the IL-76 from the moment it landed in Churchill to the point when it touched down in a region of Russia known for organized crime activity. Whether they let the Russians do what they did for intelligence purposes, or whether they were powerless to intervene, no one will say. Churchill is small and does not have immigration, security or customs people permanently stationed there.

The broader significance of the Churchill incident came into sharper focus for Leblanc the following year when a Chinese research ship, ostensibly en route to the North Pole to study climate change, rounded the coast of Alaska before getting trapped in ice. With the help of scientists at an Alaska research station and satellite data from a Canadian ice-observing network, the ship's captain was shown a way through.

Instead of heading on toward the North Pole, the ship showed up unannounced at the Inuit community of Tuktoyaktuk in the western Arctic. When the Inuit there reported the ship's presence to Canada's RCMP, a makeshift team of police officers and civil servants working in the area was assembled to investigate and board the ship. There, they found several machine guns and an unaccounted-for passport. When questioned, the ship's captain said the crew was there to meet up with a Chinese tour guide who, intelligence officials later learned, happened to be a Chinese national who had claimed refugee status in 1993. The unlikely presence of a refugee posing as a tour guide in an Inuit community infrequently visited by tourists was suspicious at best.

When Jack Orr, the Fisheries and Oceans narwhal wrangler, reported that Inuit whale hunters had spotted a submarine in Cumberland Sound that same summer, Leblanc had a new appreciation for the weaknesses in the Arctic security system for which he was largely responsible. If Canada's back door was vulnerable to suspicious entries such as these, he wondered, what would the situation be like in twenty or thirty years' time when climate change was expected to melt sea ice sufficiently and allow for partial or full passage through its Arctic waterways?

Would the military or the Canadian Coast Guard be able to stop a rogue ship if it took a run through the Northwest Passage to save 6,500 nautical miles[1] or maybe load up on freshwater from an Arctic river or lake? Would Transport or Environment Canada officials be able to clean up an oil or fuel spill if a tanker such as the *Exxon Valdez* was damaged by ice and spilled its cargo? And what about a ship that might be trying to smuggle in people? In Leblanc's mind, security and sovereignty went hand in hand with environmental protection and cultural integrity in the North.

To answer those questions, Leblanc enlisted the support of several colleagues in other government departments. The reaction was swift and positive. In the spring of 1999, representatives from the military, the RCMP, the Coast Guard, Canadian Security and Intelligence Service, Foreign

1. One international nautical mile is equivalent to 1,852 meters or 1.5 statute miles.

Affairs, Revenue Canada and Immigration met in Yellowknife for what would become the first of several biennial meetings of the Arctic Security Interdepartmental Working Group. By the time the third meeting was held in Iqaluit in 2000, several other departments, including Environment Canada, Transport Canada and Health Canada, had joined in.

The group's work took on a new sense of urgency on September 11, 2001. In the immediate aftermath of that catastrophe, virtually everything, including helicopters flying biologists around on Arctic wildlife surveys, was grounded. Everyone knew then that the world would never be the same and that North America was now vulnerable on many fronts.

The Canadian Forces moved quickly to strengthen weakness on the Arctic front. The following summer, two maritime coastal defense vessels were sent to Baffin Island to join up with land and air units. Operation Narwhal represented the first time a large joint forces exercise was held in the Arctic since the 1970s.

Around the same time that the military, Coast Guard and RCMP were coordinating these exercises at Resolution Island, another event caught Canadian officials completely off-guard. A Danish frigate was crossing the north Atlantic to replace a cairn and flag their fellow countryman had planted on Hans Island in 1988. The Danes were coming back to restake a piece of the Arctic that Canadians had assumed was theirs.

Hans Island is a treeless, kidney-shaped lump of rock that is only about 6 square miles (16 sq. km) in size. It is of little importance to Arctic birds or any other form of wildlife. Dome Petroleum briefly visited the island in the 1970s but found no evidence of oil and gas.

Although seemingly insignificant, the planting of the Danish flag on Hans Island became another lightning rod for Canadian nationalists, security experts and those caught up in the Arctic sovereignty debate. Many were outraged that a small country such as Denmark could slip in the back door and claim part of Canada as their own. Foreign Affairs officials called for calm, rightly pointing out that there were diplomatic and legal means of resolving disputes such as this.

The Liberal government under Prime Minister Paul Martin, however, was not interested in leaving it to be resolved by closed-door diplomatic chitchat. Deeply embarrassed by what the Danes had done, the government sent a few soldiers and Defence Minister Bill Graham to Hans Island in 2005 to plant the Canadian flag and reclaim the island for the nation.

The diplomatic tug-of-war that ensued spilled over into chat rooms and websites. Some of the sites mercilessly satirized the dispute between two notoriously peaceful countries sparring over a chunk of seemingly worthless rock encased in ice for as long as twelve months. So-called groups such as the Hans Island Liberation Front and Radio-Free Hans Island surfaced. A Swedish group suggested that its government should stake its own claim by erecting a statue of a popular Swedish television personality on the island. At one point in the debate, a Vancouver-based geologist threatened to go to Hans Island to stake his own mineral claim. He backed off only after Foreign Affairs officials asked him not to inflame the situation any further.

Yet the issue refused to go away. Just when Danish and Canadian authorities thought they had a protocol in place for resolving the dispute in the spring of 2008, Hans Enoksen, the Aboriginal Greenlandic leader of the Danish territory, stepped in unexpectedly, claiming that the island belonged to the Inuit of Greenland.

No one has ever seriously suggested that Canada and Denmark do battle over Hans Island. What the Hans Island situation *did* do is demonstrate how the smallest and seemingly insignificant dispute has the potential to become a much bigger issue than it should be. This was the point that Joseph Spears, a maritime lawyer who advises government and industry on ocean law issues, tried to drive home during his presentation in Halifax.

The world doesn't necessarily care what Canada thinks when it comes to controlling the Northwest Passage, he said, describing the Arctic route as a "speed bump" in the world's shipping routes. Once climate change melts the ice and turns the Northwest Passage into a viable connection between the Atlantic and the Pacific, he warned, other nations will show up in order to save time and reduce fuel costs. Given how quickly the ice is melting in the Arctic, the future of the Arctic is uncertain, he added. "The rulebook is being rewritten."

What the rules will look like in the Arctic is far from clear. There are eight countries—the United States, Russia, Canada, Norway, Sweden, Finland, Denmark (Greenland) and Iceland—that have legitimate claims to areas within the Arctic Circle. Each controls a 200-nautical-mile economic zone along its coasts. Only five of these countries—the United States, Canada, Russia, Norway and Denmark—have given any indication that they will exercise the option of claiming territory that lies beyond these boundaries. Currently, that territory, which is all underwater, is supervised by the International Seabed Authority, an autonomous organization that was established under the 1982 UN Convention on the Law of the Sea. By rules of the convention, no country has a right to extend the boundaries beyond the 200-nautical-mile limit unless they can prove that the zones of expansion are part of an undersea continental shelf connected to their territory.

With the potential to increase the size of Canada by the area now covered by the three prairie provinces, the redrawing of the map of the Arctic beyond the 200-nautical-mile zone has enormous implications. But it is just one of a number of challenges that Arctic nations face. Several other boundary disputes, including Hans Island, still need to be resolved. In the western Arctic, the United States and Canada have yet to agree on a maritime boundary line extending from the Alaska–Yukon border. The boundary line between Alaska and Siberia was supposed to have been resolved in 1990 when the United States and the Soviet Union signed an agreement to divide the area. The Russian parliament, however, has refused to ratify it. Parliamentarians claim that the deal robbed them of 30,000 square miles (78,000 sq. km) of territory. And then there is the Northwest Passage, which Canada claims as part of its territorial waters. That claim is disputed by both the United States and the European Union. As a consequence, commercial transportation companies may someday challenge Canada's claim over this area by sending a ship through the Northwest Passage.

Most countries have been content to settle border disputes among themselves peaceably or to leave it to the twenty-one-member UN technical committee that oversees the Convention on the Law of the Sea to review and recommend which country owns what in the Arctic. In dealing with territory beyond the 200-nautical-mile limit, the committee makes its recommendation on the merits of geological data submitted by each nation. The deadline for doing so depends on when a submitting country ratifies the Convention on the Law of the Sea. Ten years is currently the time a signing country has to make submission.

The planting of a Russian flag at the North Pole notwithstanding, the UN Convention on the Law of the Sea remains the best way of resolving the boundary issues beyond the 200-nautical-mile economic zone. The redrawing of that part of the Arctic map, however, will take some time. Russia has until 2009 to complete its case; Canada and Denmark have until 2013 and 2014 respectively. No one knows when the U.S. claim is going to be resolved because the Americans have still not ratified the UN Convention on the Law of the Sea. A small group of American senators championing deep-sea mining interests has blocked legislation required for the United States to go ahead with ratification.

Whether the Russians are willing to wait another decade or more to exploit what they think is rightfully theirs remains to be seen. That they would provoke a military confrontation seems doubtful. By not ratifying the UN Convention on the Law of the Sea, the United States cannot legitimately lay claim to those resources off its 200-nautical-mile economic zone. Nor can it counter those competing claims by Russia and Norway. Sitting on the sidelines also means that the United States does not have a seat on the UN commission that will eventually make recommendations on those claims.

Even if the United States were to ratify the convention, as the heads of the U.S. Coast Guard, the U.S. Navy, many in corporate America and environmentalists insist, its ability to support and assert its claims will be handcuffed by the condition of its small fleet of icebreakers. The U.S. Navy, which has the broadest maritime reach of any in the world, is almost incapable of operating in the Arctic Ocean.

In the absence of any agreed-upon Arctic border, international interest and activity has accelerated in the polar world. In addition to the

Russians planting a flag at the North Pole and the Danes planting theirs on Hans Island, the United States and Great Britain have been sending submarines beneath the ice. The Japanese are financing gas hydrate research in the Mackenzie Delta, the South Koreans are building icebreakers, the Chinese are getting heavily involved in polar research and the Norwegians are building an Arctic navy. Even the Strategic Policy Division of Australia's Defence Department was warning of the potential for global conflict over the Arctic as melting ice allowed for access to the region's oil and gas deposits.

Two things account for this frenzied interest: the world's insatiable thirst for new energy and nonrenewable resources and the prospect that the Northwest Passage will soon be navigable for a significant part of the year.

Up until recently, most of the energy reserves in the Arctic have been too expensive, too risky or too inaccessible to exploit. The rising price of oil will eventually take care of the high costs, new drilling technologies are making Arctic oil and gas more accessible and the melting ice is now quickly taking care of some of the risks.

How much oil and gas there is in the Arctic is unknown, but virtually every expert who has studied the potential agrees that the reserves are enormous. Russia currently owns the bulk of gas resources in the Barents Sea, including the massive Shtokman field, which is the largest offshore gas reservoir in the world. The region has proven reserves of 134 trillion cubic feet (3.8 trillion cu. m), an amount that could supply all of Europe's needs for seven years. The Russian Ministry of Natural Resources calculates that the Arctic territory it now controls contains as much as 586 billion barrels of oil. By comparison, all of Saudi Arabia's current proven oil reserves amount to 260 billion barrels.

A lot more energy is to be found. After an exhaustive four-year review of all the geological data available, the U.S. Geological Survey published a 2008 report that estimated there to be 90 billion barrels of undiscovered, technically recoverable oil in the Arctic, 1.67 quadrillion cubic feet (47 trillion cu. m) of technically recoverable natural gas and 44 billion barrels of technically recoverable natural gas liquids in twenty-five geologically defined areas thought to have potential for petroleum. That would account for about 22 percent of the undiscovered, technically recoverable resources in the world. Put another way, the Arctic holds about 13 percent of the undiscovered oil, 30 percent of the undiscovered natural gas and 20 percent of the undiscovered natural gas liquids in the world. About 84 percent of the estimated resources are expected to occur offshore. Given the current rate of world consumption of oil, untapped reserves in the Arctic could meet global demand for at least three years. The findings are of particular importance to the United States. The U.S. Geological Survey estimates that one-third of the undiscovered oil is located off the coast of Alaska.

The energy potential in the Arctic could be much higher than the U.S. Geological Survey estimates because the geologists who did the report didn't take into account gas hydrates frozen in the permafrost. This is the resource that has captured the interest of the Japanese and Canadian governments. Gas hydrates are found on both the Atlantic and Pacific coasts of North America and in many others parts of the world, but those located in the Arctic are among the most accessible and potentially the most economical to exploit. In November 2008, an expert panel convened by the Council of Canadian Academies conservatively estimated that the amount of methane hydrates are potentially one or more orders of magnitude larger than conventional reserves.

Energy companies in North America have already recognized the Arctic's energy potential. So far, according to the U.S. Geological Survey, energy exploration has resulted in the discovery of more than four hundred oil and gas fields north of the Arctic Circle. These fields account for approximately 40 billion barrels of oil, more than 1.1 quadrillion cubic feet (28 trillion cu. m) of gas, and 8.5 billion barrels of natural gas liquids.

The true value of this Arctic energy came to light just weeks before the Halifax conference when Royal Dutch Shell paid an astounding $2.1 billion to acquire leases in Alaska's Chukchi Sea. ConocoPhillips spent $506 million in the same lease sale. They weren't the only ones banking on an Arctic future. BP Exploration Ltd. set a record in Canada by spending $1.2 billion for three of five oil and gas leases that the Canadian government auctioned off. The leases cover about 1.5 million acres (611,000 ha) of the Beaufort seabed. BP's largest bid was $1.18 billion for a 500,000-acre (202,380 ha) parcel. The amount BP paid surpasses the record $585 million that Imperial Oil and ExxonMobil Canada paid the year before.

Energy development isn't the only thing that would benefit from a seasonally ice-free Northwest Passage. An Arctic route from either Europe to Asia or from Alaska to the Eastern Seaboard is up to 6,500 nautical miles, or 40 percent, shorter than a trip through the Suez or Panama canals. The opening of these shipping lanes would be a boon to shipping companies that move 90 percent of the world trade. Increasingly these companies are struggling to deal with piracy, higher tariffs and rising fuel and labor costs.

The opening of these shipping lanes would also be a boon to mining companies that have had their eyes on all the gold, silver, zinc, iron, diamonds and other mineral deposits buried in the more inaccessible areas of the Arctic world. The Slave Geologic Province, which is located in between and around Great Bear Lake and Great Slave Lake in northern Canada, is widely regarded as the richest untapped-mineral deposit in the world. Currently, three of the world's most productive diamond mines are operating in and around this area. Most experts believe the region's potential has barely been tapped because there is no all-weather road or any port connecting it to the outside world. Currently, all three diamond mines are using ice roads in the deepest, darkest months of an Arctic winter to get their fuel and supplies in and their unprocessed diamonds out.

Theoretically, there are only five practicable shipping routes that could get these energy and mineral resources out of the Arctic. Up until very recently, most of them were impenetrable. But the waters in the Northwest Passage are heating up. Since 1998, the number of vessels sailing through the Northwest Passage has been steadily increasing. Between 2000 and 2004, there were twenty-four vessels that made the voyage, almost twice as many as in the same period a decade before. In 2007, eighty-six ships had entered the Arctic waters along the Northwest Passage. Eleven of those made the full transect.

While the Northern Sea Route over northern Russia is now navigable for some of the summer months, experts are still debating when and whether the Northwest Passage will become safe for navigation on a seasonal basis. There is still a great deal of ice in the North American Arctic because the islands of the archipelago shade and protect the ice from the disintegrating forces of sun, winds and tides. Predictions for a seasonally ice-free Northwest Passage vary anywhere from six to fifty years. Even then, there is some doubt about the ability of some ships to get through multi-year ice that will inevitably calve off the permanent ice pack closer to the North Pole before drifting south.

There is no debating that the number of partial and full transits through the Northwest Passage will increase exponentially as oil and gas, mining and tourism activities in the Arctic heat up. The Russians are further ahead than anyone else in planning for this to happen. Part of their foresight arises from experience and the fact that the Northern Sea Route has opened up much more quickly to shipping than the Northwest Passage. The country currently operates fourteen to eighteen icebreakers (the numbers depend on which can be counted as still in service). One of them, the *Kapitan Khlebnikov,* has transited the Northwest Passage twelve times.

Russia has also been quicker than any other nation in making sure that a new generation of icebreakers, drilling platforms and tankers are built to withstand the hazards of sailing through and operating in these dangerous waters. The country is currently doing this by blending some Communist-era big-spending strategies with free enterprise principles. Recognizing that its private sector and shipbuilding yards did not have the resources or capacity to revamp the countries aging Arctic fleet, President Vladimir Putin signed a decree in 2007 establishing the United Shipbuilding Corporation. The government-sponsored project is using the state's financial power to bolster the shipbuilding sector so that it can exploit the hydrocarbons buried beneath the seabed of the Arctic. In addition to new icebreakers, the Russians plan to build forty ice-resistant oil platforms and fourteen offshore gas platforms by 2030. They are also seeking to acquire fifty-five ice-resistant tankers and storage tankers, as well as twenty gas carriers that will be capable of delivering fuel to Russia and foreign nations such as China that have already agreed in principle to finance some of these endeavors.

The region around the North Pole isn't the only area of the Arctic that the Russians are targeting. They are also pushing the concept of an Arctic Bridge that would connect the Russian port of Murmansk to the small Canadian port of Churchill in southern Hudson Bay. The Russians propose to use icebreakers to clear the way for freighters to carry oil, wheat, fertilizer and other goods to and from North America and Europe and Asia via a sea route over the North Pole. This idea has the support of both China and India and the American company that owns and operates the port of Churchill. Canada is also interested but not yet committed. To prove that it can be done, the Russians used the route to ship a load of fertilizer to Churchill in October 2007. That ship left Canada with a load of wheat.

By contrast, Canada is very far behind in preparing for the future. The Canadian Navy has not operated an icebreaker in the Arctic since the 1950s. It currently has no capacity to enter any Arctic waters that are significantly covered in ice. Most of the five icebreakers being used by the Canadian Coast Guard in the Arctic are nearing their end of life.

Canada may not be able to rely on the Americans to help it out as it did for so many decades. Two of the United States' three icebreakers, the *Polar Sea*, which is out of service, and the *Polar Star*, have already exceeded their intended thirty-year life service. The third ship, the *Healy*, is used primarily for scientific purposes. The United States' ability to operate in Arctic waters is so diminished that Admiral Thad Allen, the commandant of the Coast Guard, testified that the "nation is at a crossroads with Coast Guard domestic and international icebreaking capabilities."

Now that the Russians and the energy companies are making their move in the Arctic, Rob Huebert and Pierre Leblanc are no longer alone in wondering about threats to security, sovereignty and the environment. Naval and maritime operations in the Arctic have already been the subject of two major forums sponsored by the U.S. National Oceanic and Atmospheric Administration. The Center for Naval Analyses, a private consultant to the U.S. government, warned in 2007 that geopolitical upheaval caused by climate change could create new havens for terrorists, trigger waves of illegal immigration and disrupt oil supplies. In a report commissioned by the center, retired admiral Donald Pilling, the former vice-chief of U.S. naval operations, reiterated the long-standing view that neither Canada nor the United States has the military capability to handle future threats in the Northwest Passage.

Warnings such as these, of course, often fall flat because they rarely come with real-life scenarios that highlight the dangers. Retired Canadian colonel Gary Rice took care of that in the summer of 2007 when he created four highly credible scenarios to show how badly things could go wrong in the Arctic in the coming years as shipping lanes and polar flyways open up, and as oil and gas activity intensifies.

The first of these four scenarios involves a Chinese passenger/cargo plane flying along one of the new polar routes currently being contemplated to facilitate shorter flights between North America, Europe and Asia. The plane, traveling from Boston to Hong Kong, is forced to make an emergency landing on a gravel landing strip in the small Inuit community of Resolute (pop. 200). No one is killed when the tires blow and the plane careens off the runway, but 176 people are injured. There are only two RCMP officers and one community health nurse stationed in the hamlet. The nearest hospital is more than 1,000 miles (1,600 km) away. An observant cargo handler at the airport finds a cardbox with a "shippers Declaration for Dangerous Goods." Inside is 1 fluid ounce (25 mL) of highly infectious Mycobacterium tuberculosis.

What can Canada do?

Another scenario describes how a French nuclear submarine runs into trouble on a covert undersea voyage beneath the Arctic ice cap. (This would not be out of the question—Canadian intelligence officials suspect it was a French submarine that was spotted off the coast of Baffin Island by Inuit whale hunters in 1998.) Ice conditions in the Lincoln Sea in between northwest Greenland and the north end of Ellesmere Island make it highly unlikely for a ship to come to the rescue if the situation gets out of control. (This is also not so unlikely in light of the fact that two British sailors were killed in March 2005 in an accident on the nuclear submarine HMS *Tireless*. The submarine was on a joint United States–United Kingdom operation under the Arctic ice off Alaska when the tragedy occurred.) The French president decides not to call on NATO. He simply informs the Canadian government.

What does Canada do?

In a third scenario, the $16-billion Mackenzie Gas pipeline is completed and already pumping natural gas to markets in southern Canada and the United States. Young Aboriginal people have become increasingly militant because land claims and other governance issues have not been resolved. Nor have their people benefited from the wealth of gas being shipped south. So a group of them, unhappy with the status quo, forms the First Nations Liberation Movement. They acquaint themselves in the art of irregular warfare and start to blow up compressors and pipeline sections. The recently completed Arctic Highway connecting Tuktoyaktuk to the south is blocked.

How does Canada restore peace and order?

The final scenario is perhaps the most prescient considering Russia's stated intentions for an Arctic Bridge. In this case, the Russian Arctic container *Norsk Nova* has set sail from Murmansk to Churchill by way of the North Pole route. Three weeks before the departure, a Chechen terrorist cell affiliated with al-Qaeda bribes the poorly paid and drug-addicted commander of a badly secured storage unit at the Sevmorput naval shipyard near Murmansk. He turns a blind eye to a piece of cargo that contains a trunk-sized, man-portable, low-yield nuclear device. The theft of this nuclear device was not discovered by the Russian Federal Security Service and therefore not reported to the International Atomic Energy Agency. The terrorists' plan is to have the stolen device delivered to fellow terrorists visiting Chicago.

Four days out of Churchill, the captain of the freighter complies with Transport Canada regulations and forwards details of the cargo and his route. Since the port of Churchill is not classified as a major port, it has no radiation detection devices to screen the container once it arrives. And because the owners of the vessel are seen as trusted shippers, the Canadian Border Services Agency carries out a simple pro forma inspection.

The ship's cargo is then loaded onto southbound rail. The nuclear device spontaneously detonates, inflicting a 1-kiloton (5 TJ) ground-level explosion. All matter within 500 feet (150 m) of the fireball is vaporized. In the 1,640-foot (500 m) area outside that perimeter, winds of more than 150 miles per hour (240 km/h) destroy all the buildings in the area. Two ships, including an oil tanker, are badly damaged. Oil spills into the sub-Arctic waters of Hudson Bay.

Everyone within 3,600 feet (1,100 m) of the blast is hit with a neutron and gamma ray dose that will kill them within thirty days. Those living within 2 square miles (5 sq. km) of the blast will suffer the same fate if they remain in the area for forty-eight hours.

These four scenarios demonstrate how complicated it is going to be for one nation such as Canada or even the United States to maintain security and sovereignty in the Arctic. Complex problems require complex solutions. Patrol boats and a new icebreaker aren't going to be enough. The inevitable "scramble for control of the sea," said Martin Murphy, the research fellow at the Corbett Centre for Maritime Policy Studies at King's College in London, "will require the navy, the police and the Coast Guard to be involved. Maintaining security in the Arctic will not be a one-trick pony."

In a prizewinning essay that was written while he was at the U.S. Navy War College in Newport, Rhode Island, Canadian naval commander Scott E.G. Bishop added scientists to that list of institutions and people playing a role in the Arctic. "Not only is there a continued need for scientific research in the Arctic," he wrote, "it remains an important means of demonstrating Canadian control over its Arctic waters. Although some experts suggest that the Northwest Passage may soon be navigable, much of the passage remains poorly surveyed. Scientific research in areas such as oceanography, bathymetry, bottom mapping, climate research, marine biology and other fields is a practical manifestation of sovereignty and will make a valuable contribution to Canada's case for sovereignty over its northern water areas. The information garnered by such research is also of considerable interest to navies."[2]

Bringing order to the Arctic Ocean is not the just the mantra of military and security experts. As Arctic nations, energy, mining, commercial fishing, aquaculture and shipping companies stake their claims in northern waters, marine scientists, ocean conservation groups and Inuit leaders such as Mary Simon, Canadian ambassador for Circumpolar Affairs between 1994 and 2003, are also calling for some orderly means (which include Inuit involvement), to maintain the environmental and cultural integrity of the region. Without environmental and cultural security, they say, there is no security. And where there is no security, there is also no meaningful sovereignty.

The relevance of this argument came to light when I talked to Gary Sergy, an Environment Canada oil spill specialist, to see what he thought about Canada's ability to respond to a worst-case scenario, in which a single-hull oil tanker makes a run through the Northwest Passage, claiming, as the United States and the European Union does, that this in an international strait.

Once in Lancaster Sound, the tanker runs into a sheet of thick ice that punctures the ship's hull. Oil begins to spill out. Several thousands of narwhals and beluga whales are in the area feeding on arctic cod that are, in turn, feeding on the krill that thrive along the edges of the ice. At least two dozen polar bears are on this sheet of ice hunting seals.

Canada's flagship icebreaker, the *Louis St. Laurent*, has just left the port outside of Halifax and has run into the same computer problems that shut its engines down for four days in the Beaufort Sea during the summer of 2006. (The computer system is so antiquated that there is almost no one left in the industry that knows how to repair it.) The *Sir Wilfrid Laurier*, another Coast Guard icebreaker, is in the Beaufort helping scientists map the seabed. It is at least five, maybe six days away. The one U.S. icebreaker still in operation is nowhere near the area.

---

2. Bruce S. Oland Essay Competition – Second Prize Essay, 2008.

There are only four helicopters operating out of Resolute, the nearest air base. One is down due to mechanical problems. The other three have all been dispatched by the Polar Continental Shelf Project to serve scientists in the field. Weather issues, which are common in the High Arctic, have them grounded for the time being. What little airpower the Canadian military has at Eureka on Ellesmere Island is not enough.

Down south, both the United States and Canada are suffering through one of the worst forest fire seasons ever. Virtually every non-military plane and helicopter in North America has been seconded to deal with these infernos. (This actually happened in 2003 when fires in British Columbia, Alberta and Alaska taxed the firefighting system to the maximum.)

The amount of oil spilling out of the tanker is substantial. It flows under the ice. Strong winds and currents carry it to bird-nesting islands and the Greenlandic turbot fishery in Baffin Bay and Davis Strait. Thick fog and cloud cover make it impossible for either satellites or planes to track the flow of oil, as it is eventually separates from the ice. The *Exxon Valdez* is no longer North America's worst man-made disaster.

Sergy was part of a pioneering group that conducted oil spill experiments in the Arctic in 1981. As much as was learned back then, Sergy told me, policies and priorities changed and the emphasis shifted away from the North. Canada today is ill prepared to face the challenges posed by such a large oil spill in a region as vast and isolated as the Arctic, according to Sergy. Yet, he says, Canada can improve before it's too late.

"We need to address where we have gaps in knowledge, people, equipment and process. For example, 'what if' scenarios need to be thought through so that strategic and tactical plans can be developed to deal with the oil spill conditions we will face. Critical environmental information to identify those areas of the Arctic that are of environmental, cultural and economic importance needs to be assembled in a form to allow rapid decision-making."

Sergy was also one of many experts brought in to help in the cleanup of the *Exxon Valdez*. If anything was learned from that disaster, he told me, it's that no amount of last-minute money can clean up a spill of that size in an appropriate manner. The best and most effective way of dealing with a catastrophe like that one, he said, is to be ready before the event. "You must preplan how to take the correct actions in the midst of initial chaos. You must first target those areas that are most important. A cleanup would have to happen very quickly."

Sergy acknowledges that lessons have been learned since the *Exxon Valdez*. But he questions how Canada can effectively deal with a spill of that magnitude in the Canadian Arctic today or in the near future: "How would you get a cleanup crew on-site with no port or airstrip? Unless you're near an air base at Resolute, you can't just get on some helicopters as they did in Alaska to deal with the *Exxon Valdez*. You need to respond relatively quickly. You need to rotate a workforce. You need heavy equipment and the ability to move everything relatively easily. You have oily waste to handle. It all boils down to a logistical nightmare. We just don't have the infrastructure. On top of this, add on the difficulty and hazard of working in extreme weather possibilities and under dangerous conditions."

The rapid response required for such an ecological disaster would almost certainly be more than one country could handle, be that country Canada, the United States, Norway, Denmark or Russia. What's required is an international mechanism by which shipping lanes, oil and gas developments, deep-sea mining, commercial fishing, sovereignty claims and other issues can be managed in a way that prevents or mitigates a disaster such as this one.

Here, there are lessons to be learned from Antarctica. Following the International Geophysical Year in 1958, polar scientists tried to build on the spirit of cooperation that made the event such a success. What followed in relatively rapid order was a treaty signed by twelve countries in 1961 and another forty-six countries that acceded to it years later. In addition to prohibiting military activities, the disposal of radioactive wastes and any economic exploitation of the continent, the treaty compels participating countries to meet regularly to discuss issues of mutual concern and resolve potential conflicts. While not perfect, it has functioned extremely well.

The idea of drafting a wider treaty to deal with Arctic issues is nothing new. University of Toronto political scientist Franklyn Griffiths came up with a proposal in 1979 that would have set up a demilitarized zone in the Arctic in which polar nations would cooperate in areas of pollution control and scientific study. Lincoln Bloomfield, the former director of Global Issues for the National Security Council in the United States, expanded on that idea with a much broader proposal two years later.

The idea of an Arctic treaty has lived on ever since, but it never really caught on seriously until the spring of 2008 when Scott G. Borgerson, a fellow at the Council of Foreign Relations, warned in the influential journal *Foreign Affairs* that the United States cannot afford "to stand idly by" and watch as events unfold in the polar world.

"The Arctic region is not currently governed by any comprehensive multilateral norms and regulations because it was never expected to become a navigable waterway or a site for large-scale commercial development," he wrote. "Decisions made by Arctic powers in the coming years will therefore profoundly shape the future of the region for decades. Without U.S. leadership to help develop diplomatic solutions to competing claims and potential conflicts, the region could erupt in an armed mad dash for its resources."

Coming in the midst of another International Polar Year, the signing of an Arctic treaty would be a fitting end to an event in which more than sixty countries participated. As much as an international agreement on managing the Arctic is needed, however, the stakes and circumstances in this case would require a very different treaty than the one currently governing lands and waters around the South Pole. Unlike Antarctica, there are people in the Arctic. Nearly 2 million people live in Russia, 650,000 in Alaska, 130,000 in Canada and a little more than 1 million in Greenland, Iceland, the Scandinavian countries and the Faeroe Islands. The interests of these people would have to be represented and accounted for in any future treaty.

It is also unrealistic to expect any country to refrain from economic activity in the Arctic given the high stakes and the investments that have been made so far. Even the Inuit of Canada, the United States,

Greenland and northern Russia would agree with this. At the very least, rules and regulations governing shipping, tariffs, drilling, disposal of wastes, safety and impacts on Aboriginal peoples and wildlife could be addressed in a more meaningful and coordinated way through the treaty process.

Rob Huebert is not alone in suggesting that the Arctic Council, which was formed in 1996 after the eight Arctic nations signed the Arctic Environmental Protection Strategy (AEPS), could facilitate a treaty such as this. In addition to the eight member nations, sanctioned observers include six non-Arctic nations—France, Germany, Netherlands, Poland, Spain and the United Kingdom—and several international organizations, including the International Union for Conservation of Nature, the International Red Cross Federation, the Nordic Council, the Northern Forum and a handful of non-governmental organizations such as the Association of World Reindeer Herders.

"The Arctic Council was created in the hopes that it would serve as an international body to facilitate cooperation between the eight Arctic nations," says Huebert. "At Canada's insistence, it also includes a role for the Aboriginal peoples of the North. What better time than now to use the council and the UN Convention on the Law of the Sea to resolve boundary issues and strengthen the rules governing the Arctic."

The human history of the Arctic is complicated, rife with human drama in sublime and picturesque landscapes, of courage and recklessness and of heartache and reward in the pursuit of riches that were never fully realized. But the last Great Ice Age is now coming to an end, and the new world is going to open the door to all that wealth that eluded so many explorers for more than five hundred years.

This end of the Arctic as we know it will likely enrich southern populations increasingly starved for energy resources. But it will also open the door to smugglers, illegal aliens and terrorists and very likely trigger a series of ecological collapses that will result in extinctions, extirpations and more species being added to the endangered list. In turn, these catastrophic changes will rattle the foundations on which Inuit, Inuvialuit, Gwich'in and Dene cultures are built.

The Arctic is no longer a no-man's land of interest only to missionaries, military strategists, outdoor adventurers and the Aboriginal peoples who live there. In the not-too-distant future, the forces of climate change are going to transform this icy world into a new economic frontier. The end of the Arctic will be the beginning of a new chapter in history. The Age of the New Arctic remains to be written.

EPILOGUE

Everywhere
A huge nowhere,
. . .
An arena
Large as Europe
Silent
Waiting the contest.

—F.R. Scott, "Flying to Fort Smith," 1967

THE FIRST SIGN OF WINTRY WEATHER that normally comes to northern Ellesmere Island in late July—a fresh dusting of snow on the Sawtooth Mountains—hadn't yet presented itself. Instead, hundreds of mosquitoes were clinging tightly to the fuzzy shell of my fleece jacket; a stiff warm breeze was the only thing stopping them from feasting at their leisure.

Sitting on the side of a hilltop on this, the most northerly island in the Arctic, I was waiting for a trio of wolves that had made a practice of coming this way late in the evening. After about two hours of soaking up the enervating warmth of the midnight sun, I got lost in studying the mountains and dozed off. And then something, not a sound or a shadow, but a feeling that I was being watched, startled me back to the alert state I should have maintained in polar bear country. When I turned around, I saw a big white wolf looking down at me as if he was seeking company.

I looked at him and he looked at me, and then just like that, the animal turned and disappeared. By the time I clambered to the top of the hill, the white wolf was long gone, lost somewhere in the deep ravine that snaked its way down to the icy shores of Eureka Sound.

This was the last of eleven trips I had made to the Arctic since I started this project on climate change. During that time with wildlife biologists, glaciologists, oceanographers, climatologists, ice physicists, disease specialists, anthropologists, lawyers, security and military experts, Aboriginal leaders and hunters, it was clear that the Arctic world was collapsing and rapidly moving into a new state. The warm weather and the sight of those three wolves with tongues hanging out from the sides of their mouths seemed a fitting and somewhat disturbing end to all that I had learned about the past, present and future climate of the Arctic world and what it means to the rest of the planet.

Warm weather is not extraordinary in Eureka, the so-called "Garden Spot of the Arctic," where one nearby creek is called "Hot Weather." But the summer heat of 2008 radiated intensely for so long in nearly every corner of the archipelago that even the most seasoned scientists were wondering if it would break the record set the previous year for the lowest ice cover since satellite data was first recorded in 1979.

Many scientists saw 2007 as the "tipping point" that climatologist Mark Serreze described to me when I started this project. That's the point when winter's freeze can no longer keep up with the summer melt. The heat experienced in 2007 affected everything from glacial runoff to sea ice thinning, the habitat of pikas to polar bears, the Arctic fishery in Alaska, the distribution of arctic cod in Hudson Bay, the continued spread of avian cholera off the coast of Southampton Island and Russia's Arctic ambitions. For the third year in a row, hundreds of arctic whales made the mistake of staying in the Arctic longer than they should have. Here on Ellesmere Island, the Inuit at Grise Fiord were stockpiling sea ice because glacial runoff was no longer providing them with enough drinking water. Farther south at Pangnirtung on Baffin Island, unusually heavy rains and melting permafrost resulted in widespread flooding, the collapse of two bridges and the need to dump raw sewage into the pristine waters of the fiord. The Nunavut government was forced to call a state of emergency.

Stunned was the adjective many scientists used to describe how they felt when they saw the satellite data that suggested the Arctic might be seasonally ice-free in as little as five to eight years instead of the thirty to fifty years as predicted only a few years ago. *Shocked* was a word frequently used by others who watched as the old multi-year ice in the Arctic virtually disappeared in some places where it has always been extensive and piled up in high-pressure ridges.

For those scientists who wanted to see how the winter of 2007–08 would unfold before passing judgment on whether the tipping point had been passed, an answer was not long in coming. Ice was the last thing ice physicist David Barber was worried about when he, Louis Fortier and an international team of scientists made plans to have their research icebreaker frozen into the Beaufort Sea that winter. When they sailed the *Amundsen* into the western Arctic in November, the ice that traditionally begins to form in late October hadn't even begun to gel. By mid-December, the southern Beaufort Sea was still wide open.

Barber and his colleagues got an even bigger surprise when they sailed farther north into McClure Strait, which is legendary for being a major barrier to transits though the Northwest Passage because of thick, multi-year ice that piles in from the Beaufort Sea. Instead of being choked with rock-hard ice tens of feet thick, the strait was completely ice-free. Back home in Winnipeg, Barber told me the eighteenth-month voyage was one of the most remarkable of his career.

The Arctic winter of 2007–08, however, ended up being so cold that climate change deniers were gleefully predicting in February that the end of global warming was at hand. Despite the frigid temperature in the mid-to-late stages of the season, it was unable to slow the momentum of the thaw that had been eating away at the ice for more than a decade.

When Andrew Derocher traveled north that spring to resume the polar bear research that he and Ian Stirling had been conducting in the Beaufort Sea, the ice retreat in the western Arctic was already well under way and far ahead of schedule. So much ice had melted the previous summer that the hard bite of winter couldn't bring it all back. Many of the polar bears that Derocher was searching for were nowhere to be found. By April 25, he had spotted just twenty bears, one-third of what he would normally expect to see. There was so much fog and open water in the area that during a particularly brutal two-week stretch of bad weather, he and postdoctoral fellow Greg Thiemann were able to fly in the helicopter for just one day. Derocher had never seen it as bad as this in the Beaufort. At the time, he feared that what he was seeing was the beginning of an ecosystem collapse. The ice that is so crucial to the welfare of polar bears, whales, seals and other marine mammals was disappearing, and it was disappearing at a rate far faster than anyone was predicting.

By the time Derocher had flown back south, it had become apparent to him, to Stirling and to many others where some of the polar bears had gone. Three of them had wandered 300 miles (500 km) inland to the southwest shores of Great Bear Lake, where they were shot by an RCMP constable who, like everyone else in the Dene community of Deline, had never expected to see a polar bear this far south. The starving animals were in such an emaciated state that he and a colleague had little trouble throwing the carcasses into his pickup truck.

The same thing happened in Alaska and northern Quebec, where two polar bears wandered hundreds of miles inland into the forest. The one in Alaska was hunted down and shot by a trapper. The other in northern Quebec was last seen at Manitou Gorge on the Caniapiscau River, where it was apparently in such a bad state that it had treed a porcupine in the hopes of getting a meal.

When considered individually, anecdotes such as these are as meaningless as claims that deniers were making about winter's deep freeze in 2008 signaling an end to climate change in the Arctic. One year of sightings of any kind isn't enough to establish a trend. But occurring as they did during a four-year period of rapidly thinning ice, drowning polar bears in the southern Beaufort Sea, increasing observations of cannibalism and starvation in polar bear populations, and greater numbers of bears in Alaska denning on land instead of ice, these reports were further indications that polar bears at the southern edges of their range could be in trouble. The sighting of nine polar bears swimming 12 miles to 62 miles (20–100 km) offshore in the Chukchi Sea a few months later seemed to add an exclamation point to that prospect. So did the blood samples that Stirling had taken from polar bears during the spring I traveled with him to launch this project. When compared to those taken from bears twenty years earlier, Seth Cherry, a Ph.D. student working with Stirling, Derocher and Evan Richardson, found that the ratio between urea and creatinine, waste materials that the polar bear recycles, was lower. That meant the bears are now fasting longer than they normally would be, which suggests that they are either having trouble finding ringed seals on thin ice or thin ice has removed the steady platform that seals need to nurse their young.

Change came so fast to the Arctic in 2007 and 2008 that some scientists were finding it increasingly risky to be in the field. It wasn't just thin ice and thick fog rising over so much open water that were hazards. Avalanches, rapidly melting glaciers and thawing permafrost were also causing the land and seascape to be unsteady.

In July 2007, glaciologist Garry Clarke and his colleagues were shutting down their camp at the Trapridge Glacier in the Yukon when a massive rock and ice avalanche came tumbling down the north face of Mount Steele. Two days later, an even larger slide on the slope registered on seismometers around the world. Mount Steele was 9 miles (15 km) away. But not knowing what was causing the slide at the time, Clarke considered the possibility of evacuating the area nonetheless.

News of the slide reverberated throughout the scientific community not only because it was so massive, but also because an earthquake had not triggered it. The most likely explanation for this and other big slides that had rocked the Yukon and Alaska in recent years was that climate change, frost shatter and permafrost degradation had destabilized the mountain slopes.

Queen's University scientist Scott Lamoureux and his colleagues were up in the High Arctic on Melville, the largest uninhabited island in the world, right around the time that slope on Mount Steele was disintegrating. Summers on the island had been left largely unaffected by the warming taking place across the Arctic. July temperatures have remained steady at about 4°Fahrenheit (5°C), partly because there is so much ice in the area. That July, however, Lamoureux and his associates basked in temperatures of 68° Fahrenheit (20°C) or more. The heat was so intense that it thawed the permafrost 3 feet (1 m) below the surface. Throughout those warm weeks, the scientists watched in amazement as the meltwater below lubricated the topsoil, causing it to slide down slopes, clearing everything in its path and thrusting up ridges at the valley bottom. The landscape piled up like a rug and was being torn to pieces. Had this happened in a populated or industrial area, Lamoureux told me later, the impact would have been catastrophic.

Glaciologist Sarah Boon had an even closer call on the Belcher Glacier of the Devon Ice Cap, where she was collaborating on a long-term study with glaciologist Martin Sharp. Boon was alone with graduate student John Padolsky when a huge avalanche of slush separated them from their camp. Boon knew they were in trouble when they poked and prodded with their poles, trying to find a spot to cross. It was 7 feet (2 m) deep in some places. In others, their poles couldn't touch bottom.

Unable to get above or below the giant slurry of slush, the pair found a spot that seemed shallow enough for them at least to try to cross. Roped together, Padolsky was the first to go. About halfway across, the waist-deep slush began to solidify around his legs.

Still on solid ice, Boon tried to pull him out with the rope but failed. She shouted to Padolsky to lean forward and take the weight off his legs, then crawl out. Knowing that he had to get out fast, Padolsky did so and freed himself.

Just as it looked like he was finally in the clear to get to camp, another glacial stream opened up, stranding him on the one high point of ice that was not surrounded by a pool of water or slushy ice. Realizing the danger they were in, Boon called Barry Hough and Mike Kristjanson at the Polar Continental Shelf Project (PCSP) in Resolute for help. Then for the next three hours she hopped around to keep warm while talking to Padolsky, telling him stories, hoping to keep him alert and upbeat—and giving PCSP hourly updates via satellite phone. Soaking wet from the waist down but huddled in a space blanket suit and several layers of down clothing, Padolsky grew increasingly colder as the sun disappeared behind a ridge.

Three hours later, a helicopter came in and pulled the pair from their precarious perch. They were lucky that the good weather at the time facilitated safe flying for a rescue.

The summer of 2008 stopped slightly short of breaking another record for low ice cover in the Arctic. Nevertheless, the U.S. Geological Survey reported in 2008 that most glaciers in every mountain range and island group in Alaska were either thinning, retreating or in a state of stagnation. Ninety-nine percent of the largest glaciers in Alaska were in retreat. Looking at land ice from satellite images, NASA reported in December 2008 that more than 2 trillion tons of landlocked ice in Alaska, Arctic Canada and Greenland had melted in five years.

The continued meltdown still wasn't enough to convince the Canadian or American governments to take action. In Canada, the government ignored sound advice from Environment Canada scientists on a number of climate change issues. Government scientists were being muzzled to the point that not one of them was allowed to attend the last two meetings of the Kyoto Protocol in Bali and Poland. Nor were they allowed to speak to the media or the public without government permission. The restrictions were so draconian that scientists with the Canadian Wildlife Service, the Meteorological Service of Canada and the National Water Research Institute were at war with the government.

Even the troubled status of the polar bear failed to inspire a plan. It took nearly two years of lobbying, court action and an exhaustive scientific review by a blue-ribbon panel of experts to finally convince the Bush administration in the United States of the need to list the polar bear as a "threatened species" in 2008 because of the predicted loss of their critical sea ice habitat as a consequence of climate warming. Yet in Canada, the scientific committee that makes recommendations on such listings hired a biologist who was highly skeptical that greenhouse gases were causing climate change to review the status of the polar bear. Not only had Mitch Taylor, Nunavut's former polar bear biologist, been critical of the U.S. Fish and Wildlife's decision to recommend listing the polar bear as threatened, he had signed the Manhattan Declaration, which claims, among other things, that controlling greenhouse gas emission is a "dangerous misallocation of intellectual capital and resources" and a costly waste of time.

Taylor's previous contention that polar bears will adapt to a warming world had not sat well with fellow polar bear scientists on the International Union for the Conservation of Nature's Polar Bear Specialist Group either. In 2005, they unanimously (including Taylor) agreed to designate the polar bear as vulnerable, which is equivalent to the threatened designation. So it was a disappointment, but no great surprise to them, when Taylor and his co-authors rejected the conclusions of the U.S. studies and made all their modeling projections of future population trends on the basis of stable sea ice habitat.

Given the risks, the harsh conditions and the time away from family and friends, many scientists I had met along the way were clearly frustrated that no one was listening to them, that the future of funding looked so gloomy and that an action plan to adapt to and face the changes that climate change posed was still not in the making. Faced with this and the impending closure of the atmospheric research laboratory on the Fosheim Peninsula near Eureka and an end to funding for the Canadian Foundation for Climate and Atmospheric Sciences, one hundred and thirty scientists signed an open letter to Prime Minister Stephen Harper, opposition leaders and provincial premiers, warning them that the dangers climate change posed to the world were even greater now, and developing more rapidly, than previously predicted when the Intergovernmental Panel on Climate Change issued its report in 2007. Neither were they alone. In 2008, both the U.S. Geological Survey and the National Oceanic and Atmospheric Administration issued dire warnings about the accelerated meltdown in the Arctic and the effect it would have on drought, storms, and rising sea levels in other parts of the world.

Despite the bad news of the past two years, there were a few reasons to be optimistic. The Canadian government committed itself to building a new icebreaker, an Arctic port and a High Arctic research station. And as much waffling as there was on climate change initiatives, sixty-three countries, including Canada, came up with special funding for scientists to participate in the International Polar Year (IPY), a scientific program that focused on a number of polar issues in the Arctic and Antarctica between 2007 and 2009. Scientists in Canada and the United States were hopeful that they could build on the momentum that came with the insights and breakthroughs made in that time. The impending end of IPY also had a number of security and sovereignty experts talking about drafting an Arctic treaty, similar to the one for Antarctica that followed the last polar year event.

Most everyone, including Queen's University scientist John Smol, knows that time is running out. Smol and Marianne Douglas have been drilling into the bottom of Arctic lakes for more than twenty years, homing in on single-celled algae that have been dead for hundreds if not thousands of years. The prevalence of one species over another signals what the climate was like in the time these diatoms lived.

Those they find 1 foot (30 cm) into the sediment are around two hundred years old and well adapted for colder climates. The younger diatoms they see near the surface are much different in nature and better suited to warmer conditions. The timing of this shift in the food web in Arctic lakes appears to be directly connected to climate change driven by human activity. The shift is also likely signaling bigger changes that are about to take place in the polar world.

Smol has been on this road before. In the 1980s, he and David Schindler, another Gerhard Herzberg Gold Medal winner, were part of a group of pioneering scientists who proved that the sulfur-spewing industrial smokestacks of the 1980s were having a devastating effect on thousands of lakes in Canada and the United States. There are a lot of similarities between the acid rain debates a quarter-century ago and the climate change challenge we face today, they both told me.

"There were many people back then who doubted there was a link between acid rain and the death of lakes," said Smol. "Once we established without a doubt that it was true, politicians were forced by public opinion and the economics of the problem to do something about it. That's where we are now with climate change. We've got to convince politicians to do what's necessary to deal with this global problem."

There was another night at Ellesmere Island when I returned to that same hilltop site waiting for the wolves to show up. The air was cold but not frigid. Nothing was moving. The descending trill of a rough-legged hawk broke the hush of this perfect night. This time I got lost studying the broken-windowpane shatter lines of aquamarine meltwater etched in the ice of Eureka Sound. In the midst of that marvelous scene, a ringed seal rose from its breathing hole, stretched itself onto a floe and went to sleep.

I had learned a great deal from scientists and Inuit hunters on those trips and was grateful for all they generously shared. But it was this simple scene that seemed to sum up all the lessons I had learned in those eighteen months. Arctic ice can be a hideous force, capable of crushing ships, stranding whales, starving migratory seabirds and wiping out those miniature beavers that were on constant lookout for the ancestral black bears, weasel-like carnivores and Eurasian badgers on this island 4.5 million years ago. Yet, for that seal I was gazing upon, it was a steady platform that it used to rest, to deliver and nurse its pups and to hide from polar bears. Ice is the greenhouse shelter in which algae blooms and krill feed. Without ice, those narwhal and belugas that Jack Orr caught in Repulse Bay and Cumberland Sound would have one less signal telling them when to come, when to leave and where to go for arctic cod and capelin. The walrus that Gabriel Nirlungayuk and I hunted near Marble Island would be forced to haul out on land rather than on ice floes floating far offshore.

Ice is what buffers those frozen shores along Tuktoyaktuk, Aklavik, Shingle Point, Herschel Island and Shismaref in Alaska from the erosive forces of autumn storms and rising sea levels. Without glacial ice and sufficient snowpack in around the Brintnell Glacier, the Nahanni would not be the legendary river that it is and the Mackenzie River would not overflow its banks and refill the forty-five thousand lakes that lie within its great delta. Neither would the mercury that is naturally stored in the frozen ground melt out and enter the food chain.

Ice is what allows Peary caribou and muskoxen to travel from one island to another in the archipelago when food is scarce. It's how those pikas in the St. Elias Icefields of the Yukon found a refuge in *nunataks*.

Ice is as much a metaphor for human struggle and triumph as it is a force of nature. Hundreds and thousands of years ago, people such as Kwaday Dän Sinchi (Long Ago Person Found) set out in search of caribou and game or perhaps new places to live and hunt. For Kwaday Dän Sinchi, it may have been that glacier in which he was entombed that led him to the valley along the Yukon–British Columbia border. For the Inuit of Nadlok on the Burnside River, on the other hand, it was the thick year-round ice on the Arctic coast that drove them inland to hunt caribou instead of seals and whales.

The human history of the Arctic is filled with thousands of stories. The notion of a northwest passage snaking through the Arctic Archipelago may have been based on primitive and flawed views of the geography of the world. The pursuit may have been driven by greed, hubris and national pride. But more often than not, the outcomes were dictated by thick Arctic ice. With the benefit of hindsight, the history of the Arctic would have been a much different one had the search for a northwest passage not taken place during the Little Ice Age.

Now that melting sea ice has trigged another race in the Arctic, the stakes are much higher. What happens in the future matters not only to culture, wildlife, the environment, security and sovereignty. It matters to the rest of the world.

ACKNOWLEDGMENTS

THE IDEA FOR THIS BOOK got its start with an article I wrote for *Equinox* magazine in 1992. "The End of Arctic" considered how climate change might affect wildlife, the environment and the people of the Arctic. Coming as it did early in the climate change watch, the article was largely speculative. The issue continued to intrigue me as events unfolded just as the scientists in the article predicted. The opportunity to pursue this subject in depth came two years later when I was awarded a Knight Science Journalism Fellowship. That gave me a year at Harvard and the Massachusetts Institute of Technology (MIT) to study the issue with some of the finest thinkers in the world. Fortunately, I was invited back in 2001 for a weeklong "boot camp" at which climate experts from Harvard, MIT and other institutions shared their thoughts and research with me and eleven other writers, journalists and documentary filmmakers.

I knew there was a book in it then. But the opportunity to start writing it didn't come until the late spring of 2006 when John Honderich, the former editor and publisher of the *Toronto Star*, called to let me know I had won the Atkinson Fellowship in Public Policy. As every journalist in Canada knows, the Atkinson is one of the most coveted journalism fellowships on the continent. Since its establishment in 1988, it has aided research into some of Canada's most important and often most complicated policy debates. I proposed to look once again at climate change and its impact on wildlife, the environment, culture, as well as the implications it has for security and sovereignty.

Given John's interest and expertise in Arctic issues—he wrote a very good book on Arctic sovereignty, *Arctic Imperative: Is Canada Losing the North?*, more than twenty years ago—I'm certain that he was the main reason why I got the fellowship. He was also chair of the selection committee. It didn't hurt that David Schindler, one of the finest scientists of our time, and Gregory Taylor, the dean of science at the University of Alberta, nominated me for the fellowship. Credit and thanks must also go to Janis Gross Stein, director of the Munk Centre for International Studies at the University of Toronto; Peter Armstrong, president of the Atkinson Charitable Foundation; and especially Charles Pascal, executive director of the Atkinson Charitable Foundation, who supported me every step of the way through the fellowship months. Glen Colbourn did a very fine job editing the original series for the *Toronto Star*. Joe Hall, the managing editor of the *Toronto Star*, and Allan Mayer, the editor of the *Edmonton Journal*, were extremely supportive and saw to it that the stories were given prominent play in those two fine newspapers. The fact that the series was chosen as one of five finalists for the $75,000 Grantham Prize, the biggest prize in environmental journalism, is as much a credit to them as it is to me.

The book covers some of the same ground. But most of what is written here will be new even to those who read the series. Of course, neither the series nor the book would have gotten anywhere had it not been for the scientists and the Inuit who allowed me to join them in the field and for the Canadian Coast Guard, which gave me the opportunity to sail on the *Louis St. Laurent* icebreaker as it made its way through the Northwest Passage in the summer of 2007. For this I thank: Ian Stirling, Andrew Derocher, Eddy Carmack, Mike Demuth, Dan McCarthy, Jack Orr, Stan Boutin, David Henry, Ben Gilbert, Scott Dallimore, Oleg Mikhailov, Josée Lefebvre, Captain Andrew McNeill, Gabriel Nirlungayuk, Davidee Evic, Robie Dialla, Tommy Qaqqasiq, Noah Mosesee, Jonah Siusangnark, Paul Tegumiar, Luky Putulik, Laurent Kringayark and Mark Tagornak. Thank you as well to Commander Kenneth P. Hansen, who invited me to participate in the Maritime Security Conference in Halifax, and to John Gunter of Tundra Buggy Tours and Robert Buchanan of Polar Bear

International, who hosted me in Churchill, Manitoba. I am indebted to Pierre Leblanc, the former commander of the Northern Forces, who set aside a considerable amount of time to share his experiences and thoughts on sovereignty and security.

I learned a great deal from other scientists in the field as well. Oceanographers Bill Li, Sarah Zimmerman Vlad Kostylev and Ed Hendryks were terrific companions and wonderful instructors on the *Louis St. Laurent*. On two occasions, once in the Yukon and another time at the Butterfly Glacier in the Northwest Territories, I had the pleasure of spending time with Christian Zdanowicz of the Geological Survey of Canada, who was very generous in sharing his thoughts and expertise on climate change and glaciology. That didn't come as a surprise because the legendary Fritz Koerner was the first scientist I talked to when I wrote that *Equinox* article in 1992. He was a mentor to both Christian and Mike Demuth at the Geological Survey. As everyone in Arctic science knows, Fritz was not only a great scientist, he was an extremely thoughtful, generous and remarkably adventurous man. His death in the spring of 2008, shortly after his forty-ninth consecutive research season in the Arctic, was an unexpected and tragic loss, as was the death of Davidee Evic, who worked with Jack Orr and other scientists refining the art of beluga whale capture off the coast of Baffin Island.

To the northern research station directors—Marty Bergmann, the director of Canada's Polar Continental Shelf Program (PCSP), Andy and Carole Williams at the Kluane Research Station, Mike Goodyear at the Churchill Northern Studies Centre and everyone associated with the Aurora Research Institute in Inuvik and the Polar Environment Atmospheric Research Laboratory on Ellesmere Island—I thank them for the food, drink, insights and wonderful conversations they offered. PCSP's Barry Hough and Mike Kristjanson were, as usual, terrific in helping me get in and out of a number of research camps in the High Arctic. I was fortunate to have met Barry in the early 1980s when Polar Shelf was prospering under the directorship of George Hobson, who took over from the legendary Fred Roots, the first PCSP director. Marty's recent appointment to the position is a sure sign that there are still good people working at high levels in government and that the Arctic program is beginning to get back on track.

Many scientists kindly read parts of this book, offering suggestions, pointing out mistakes and answering questions. The list flatters me: Ian Stirling, Eddy Carmack, Humfrey Melling, Andrew Derocher, Ole Nielsen, Don Russell, Anne Gunn, Peter Clarkson, Ryan Danby, Greg Hare, Gary Sergy, Stan Boutin, Chris Burn, Rob Huebert and Joseph Spears all took time to help. If there are mistakes in this book, they are mine.

There were many other scientists and Aboriginal leaders who were extremely generous with their time but who do not appear in the book. They include: Steve Ferguson, Jim Reist, Pierre Richard, Sam Stephenson, Ray Alisaukas, Mike Flannigan, Tony Gaston, Susan Kutz, John Nagy, Mark Nuttall and Walter Bayha of Deline, whom I met on the trapline on the shores of Great Bear Lake back in the late 1970s and whom I didn't see again until we ran into each other in Tuktoyaktuk in 2007. That was a wonderful surprise.

Working with Robert Hickey, the multi-talented editor at John Wiley & Sons, was really a pleasure. Robert was right 99 percent of the time. Not once, however, did he ever make me think that maybe I was wrong. I was also thrilled to have Heather Sangster copyedit the book. Her razor-sharp attention to detail and felicity of style was as humbling as Robert's ability to keep me on track. Thanks as well to Elizabeth McCurdy for ushering the book through production. Most of all I would like to thank my wife, Julia, and my children, Jacob and Sigrid, for being so patient, helpful and understanding through my long absences both away from home and while I was upstairs in the office writing the book. Julia, you truly are the love of my life.

INDEX

A

C

N

O

P

T